Pump Application
Desk Book

Pump Application Desk Book

Paul N. Garay, P.E.

Published by
THE FAIRMONT PRESS, INC.
700 Indian Trail
Lilburn, GA 30247

Library of Congress Cataloging-in-Publication Data

Garay, Paul N., 1913 -
   Pump application desk book / Paul N. Garay.
      p.   cm.
   Includes index.
   1. Pumping machinery--Handbooks, manuals, etc.
I. Title.

TJ900.G37   1989   621.6--dc19   87-45343
                                        CIP
ISBN 0-88173-043-2

Published by The Fairmont Press, Inc.
700 Indian Trail
Lilburn, GA 30247

Printed in the United States of America

10 9 8 7 6 5 4 3 2 1

ISBN 0-88173-043-2   FP

ISBN 0-13-741125-1   PH

While every effort is made to provide dependable information, the publisher, authors, and editors cannot be held responsible for any errors or omissions.

Distributed by Prentice-Hall, Inc.
A division of Simon & Schuster
Englewood Cliffs, NJ 07632

Prentice-Hall International (UK) Limited, London
Prentice-Hall of Australia Pty. Limited, Sydney
Prentice-Hall Canada Inc., Toronto
Prentice-Hall Hispanoamericana, S.A., Mexico
Prentice-Hall of India Private Limited, New Delhi
Prentice-Hall of Japan, Inc., Tokyo
Simon & Schuster Asia Pte. Ltd., Singapore
Editora Prentice-Hall do Brasil, Ltda., Rio de Janeiro

Contents

Chapter *Page*

Introduction

Few engineered artifacts are as essential as pumps in the development of the culture which our western civilization enjoys. From the smallest to the largest, every facet of our daily lives is served in some measure by such machines. Ancient civilizations, requiring irrigation and essential water supplies, utilized crude forms of pumps which, their design having been refined, are in use even today. Moreover, in today's precise mechanical environment, the types and forms of pump equipment add hundreds of variations to the earlier, simple forms.

The large number of forms and types of pump equipment in common use have different purposes and varying application requirements. This text has been prepared to bring together necessary information for those who are required to select and apply pumps in systems for all kinds of fluids and purposes. It is not a design manual. It limits the discussion of design to those factors which are necessary for an understanding of pump operation. The book contains not only descriptive information of many types of designs, but also the effect of design variation on use, economy, and reliability. Systems, an integral part of application, are discussed, so that the user of such equipment may properly select and install machines to reliably and economically satisfy his requirements. Reading of appropriate sections of the text will indicate possible problems to be avoided, and economics to be gained.

The engineering design of pumps is a science; their proper application is an art. The material contained here is a compilation of the results of extensive experience. Most of it is readily available in the

1

literature, should the reader care to investigate thousands of pages of subject matter. Some of the material is not new, as old-time pump users will realize. Much of the material is new, reflecting recent developments in the field of design and application. We hope that this text will be a ready reference for both old and new aspects of pump application system design.

For convenience, material is contained in the text referring to factors which affect flow systems: viscosity, temperature effects, common fluids, and others. Some nomographs and tables are included for easy variables. Finally, there are included some brief descriptions of equipment and their sources which are unusual enough to merit your attention.

In closing, it should be understood that the many specific references to trade names and manufacturers are not meant to imply that these names represent sole sources, or are being specifically recommended. It would be impossible to list all the various names, as almost everything being referenced is available from several sources. The references should be taken as being typical.

1

Codes and Standards

PUMP DESIGN

The design of pumps has been largely standardized so that by citing an appropriate standard, the purchaser will receive a machine designed to the strictures of that publication. Although it is possible to specify a standard manufacturer's construction and still receive a usable piece of equipment, in many cases the purchaser may prefer to specify the details of construction. As the specifications become more and more detailed, it becomes apparent that the overall design will approach that required by a standard.

The most common such standards are:

American Petroleum Institute, (API) Standard 610, Centrifugal Pumps for Refinery Services

American Waterworks Association, (AWWA) E 101, Deep Well Vertical Turbine Pumps

Underwriters Laboratories (UL):

UL 51, Power Operated Pumps For Anhydrous Ammonia and LP Gas;

UL 343, Pumps For Oil Burning Appliances

UL 1081, Swimming Pool Pumps, etc;

UL 448, Pumps For Fire Protection Service;
UL 1247, Centrifugal Fire Pumps Driven By Diesel Engines.
National Fire Protection Agency (NFPA), NFPA 20, Centrifugal
Fire Pumps
American Society of Mechanical Engineers (ASME):
ASME PTC 18.1, Pumping Mode of Pump/Turbines;
ASME B 73.1, Horizontal End Suction Centrifugal Pumps for
Chemical Process;
ASME B 73.2, Vertical In-Line Centrifugal Pumps for Chemical Process.

The American National Standards Institute (ANSI), issues these standards under their aegis. In the ANSI catalogue the standards identification numbers are preceeded by ANSI, thus: ANSI/UL 448, or it may be simply shown as ANSI 448.

The list given above is not necessarily the total compilation of available standards, but are those which are of most interest.

The features of the ANSI B73 designed pumps which are of interest in increasing the reliability of such pumps are:

Choice of open or enclosed impellers;
Dual volute in some sizes;
External axial impeller adjustment;
Replaceable casing ring adjustment for enclosed impellers;
Casing shroud plate for open impellers;
Replaceable hook type sleeve with O-ring seal between shaft
sleeve and impeller;
Dry rabbet fit construction;
Heavy duty shaft system—thick shaft, short bearing span,
short impeller overhang;
Choice of seal and gland options.

The above features are shown by most manufacturers' standard pump literature, and are typical of pumps manufactured to this standard, although details may vary from one manufacturer to the next.

A relatively new category is UMD 55, which signifies an upgraded medium-duty pump. The ANSI to UMD 55 upgrade is accomplished as discussed in the final paragraph of this chapter. Some recent studies suggest that the MTBF—mean time between failures—for

ANSI standard pumps may be only 13 months. The MTBF for the upgraded pumps may be as high as 42 months.

The API standard design pumps are the most expensive of any type of construction, although there are units which may be virtually custom built, and which may be required for specific applications. API pumps are selected for refinery operations where cost is secondary to reliability. This standard requires the purchaser to specify certain details and features. Also, it recognizes that the purchaser may desire to modify, delete, or amplify sections of the standard. Such modifications generally should be made by supplementing the standard, rather than by rewriting it. The API standard specifies all construction details, as follows:

Alternative designs;

Specified features for every factor of pump construction;

Materials;

Shop inspection and tests;

Preparation for shipment;

Drawings and other data required;

Proposal data required;

Appendices—data sheets, mechanical seal and piping schematics, specifications, material classes.

The API standard contains some 35 pages of detail, and therefore should be studied in detail. Even though such rigorous criteria may not be specified, the standard contains a great deal of information describing desirable pump design and construction.

Among large users and manufacturers of pumps, there has been discussion regarding enhancement of the ANSI requirements in the interests of improving mechanical reliability at modest cost. The initial, and most used scope of the ANSI specification is the dimensional standardization. An addendum drawing upon features contained in ANSI, API, and various international standards (ISO), would be a beneficial addition to the process duty line-up. Certain items should be considered to attain a new standard which may be identified as UMD 55, Up Graded Medium Duty Pump.

Runout: Shaft runout should be not more than .002 inches.

Balance: All impellers should be two plane dynamically balanced within 0.018 ounce-inches per pound of rotating mass.

Design: Double the L10 rating of the axial thrust bearings by using two, mounted back to back.

Specify a minimum L10 of 50000 hours rather than 17500 hours.

Sealing: Require compound labyrinth bearing isolators—nonwearing, noncontact. Lip seals shall not be used.

Lubrication: Synthetic diester, polyglycol or PAO synthesized lubricants shall be the preferred method of lubrication.

PUMP APPLICATION

There are many publications relating to the application of pumps to various tasks. The one most universally referred to is the Hydraulic Institute Standards. "An Institute Standard defines the product, material, process or procedure with reference to one or more of the following: nomenclature, composition, construction, dimensions, tolerances, safety, operating characteristics, performance, quality, rating, testing and service for which designed." This is a quote from the by-laws of the Institute, Section B. The publication, however, also includes much engineering information and explanatory data not falling within the classification of Institute Standards. Industry generally regards this publication as a basic reference.

Another set of specific codes refers to the selection, use, and installation of pumps dedicated to fire-fighting services. The National Fire Protection Association (NFPA), Technical Staff on Fire Pumps is made up of representatives of Underwriters Laboratories (UL), of both the U.S. and Canada, Insurance Service Offices (ISO), Factory Mutuals (FM), Industrial Risk Insurers (IRI), national trade associations, state government and engineering organizations. NFPA Standard pamphlet No. 20 is the work of these groups.

NFPA No. 20 deals with selection, and installation of pumps supplying water for private fire protection. The standard is recognized by both stock and mutual insurance organizations. The listing status of such pumps is designated by labels shown on the next page. Further information is contained in the section under "Fire Pumps."

The major differences between commercial pumps and qualified fire pumps are the standards and specifications established by UL, FM, and the NFPA. Approved fire pumps must bear the endorsement of one or more of these organizations. Approved fire pump controllers are the combined automatic and nonautomatic types, and must be built in accordance with the latest requirements of the NFPA as outlined in their pamphlet No. 20. Controllers are listed under the reexamination services of UL and are specifically approved for service by FM. They are also certified by the Canadian Standards Association (CSA), and are labeled by the Underwriters Laboratories, Canada (ULC). Unapproved pumps, of course, may be used for fire service, but cannot be labeled for fire service. Approved pumps must be used only for fire service. A prospective user should contact his insurance company to determine the correct approval agency.

2

Pump Classification

GENERAL CATEGORIES

The supply of water is one of the fundamental needs of any society, be it human or animal. The need to transfer water has therefore generated the design of a myriad of mechanical devices, which can be categorized as pumps. Except as a matter of historical interest, we need not be concerned with animal or human-power pumps. Millenium-old designs for bucket wheels and paddle devices have not, to a significant degree, been carried forward into modern technology. With the exception of open screw pumps, all of today's devices act to force water into the desired channels or pipes.

The multiplicity of today's pump design is based on the need for specific designs to suit various applications—pressure, quantity, temperature, type of fluid. It is not generally realized that the pressure of competition among manufacturers has engendered many variations to meet certain specific applications. Each manufacturer may handle a given requirement in a manner slightly different from that of a competitor. Each variation is patented, and is offered on the basis of its alleged superiority for the application in question.

Fig. 2-1 is a chart of the many varieties of pumps available for current uses. Generally, they may be divided into two basic groups— positive displacement and dynamic (kinetic). A third, less important group, is the lift pump.

8

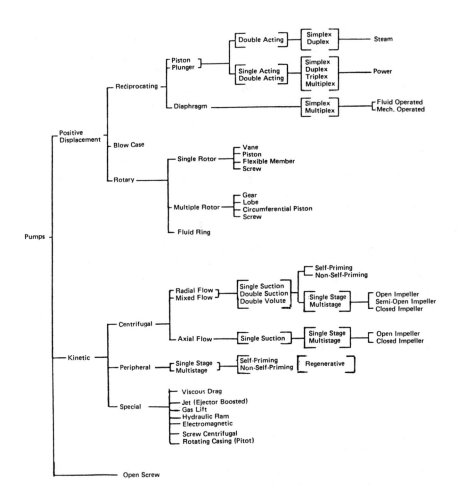

Fig. 2-1. Classification of Pumps
(Courtesy of Hydraulic Institute)

(1) **Positive Displacement:** In this category the fluid is forced to move because it is displaced by the movement of a piston, vane, screw, or roller. Positive displacement pumps act to force water into a system regardless of the resistance which may oppose the transfer.

(2) **Dynamic or kinetic:** This group of pumps has been developed in historically recent times. It may be said, however, that a

person swinging a bucket filled with water, throwing the water into an elevated receiver, is using a dynamic pumping principle. Commonly, however, when motion is imparted to a fluid—usually a rotary motion—the increase in potential energy due to the fluid motion may be converted into a higher degree of pressure, concurrent with the transfer of the fluid. This, then, is the dynamic principle of pumping.

(3) Lift, etc.: In this category are pumps which have been in use for millenia. A lift pump may be a simple arrangement of buckets attached to a vertically-moving endless chain, belt or rope. In some cases, with this same configuration utilizing paddles or sweeps, fluid may be moved essentially horizontally. This variation may be referred to as a "drag" pump. All these variations may be characterized as positive displacement, inasmuch as the quantity of fluid moved by each element is relatively fixed. This class, of course, is not capable of any pressure increase.

POSITIVE DISPLACEMENT CLASSES

Diaphragm: Fluid is transferred by the pressure of a diaphragm which flexes to form a cavity which is filled by fluid. These pumps can move virtually any fluid, are temperature-limited, and are infinitely adjustable in capacity and discharge pressure, by regulating the movement of the diaphragm. Commonly used for pumping mud and heavy slurries, the diaphragm is flexed by compressed air pressure, in one arrangement. Simply, fluid is on one side of the diaphragm, and pressurized air on the other.

A different arrangement of a diaphragm pump is designed for metering service, and where low flow and high pressure capacity are required. Typically, this design is used for pumping chemicals to a boiler, or metering strong chemicals to a water purification unit. Fig. 2-2 illustrates the construction of such pumps.

Resistance to any kind of corrosive fluids is attained by selecting the appropriate material for the diaphragm and the valves. The diaphragm, in one case, may be actuated by air pressure. In more expensive versions, mechanical linkage is used to flex the diaphragm. A

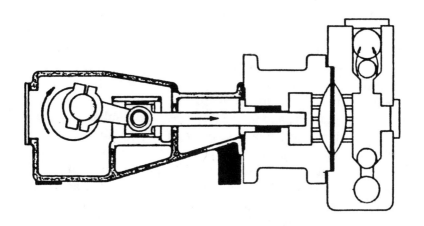

HORIZONTAL SINGLE-ACTING FLAT DIAPHRAGM PUMP

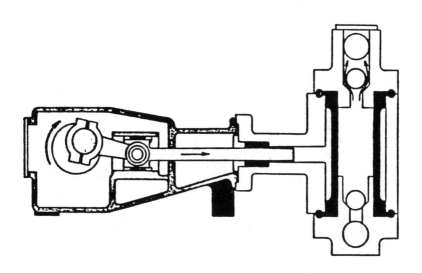

HORIZONTAL SINGLE-ACTING CYLINDRICAL DIAPHRAGM PUMP

Fig. 2-2

further degree of sophistication is possible by using two diaphragms: the first one is actuated mechanically, and pumps a hydraulic fluid; the pulsed hydraulic fluid is then used to flex the second diaphragm. With this arrangement, possible rupture of the second diaphragm, the one in contact with the pumped fluid, will not release chemicals to the outside, or to the drive system.

Reciprocating: A tight-fitting piston in a closed cylinder, or a loose-fitting plunger, acting as a displacer, are familiar versions of the common reciprocating pump. Capable of almost any pressure, and of large flow capacity, piston pumps are not as popular as they were before efficient centrifugal types dominated the market. Piston pumps in large sizes are expensive, but they have the advantages of being easily controlled by stroke adjustment or by variable speed, the ability to develop high pressures in single stage, and high reliability. Disadvantages are the necessity for slow speed operation, and a pulsed output. NPSH requirements for a reciprocating pump are more complex than for rotary pumps, due to the pulsed nature of the suction. One variation of the piston pump may be steam or pneumatically driven, utilizing a piston to do the pumping, with another piston mounted on the same piston shaft to receive the motivating steam or pneumatic pressure. Fig. 2-3 is a Worthington direct-driven feedwater pump.

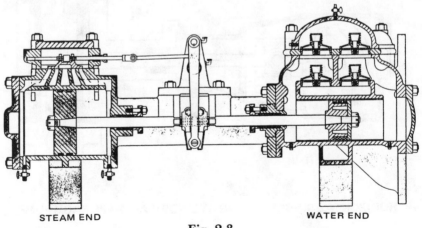

STEAM END   WATER END

Fig. 2-3

(From PUMP WORLD, Vol. 6, No. 2, 1980)

As stated earlier, positive displacement pumps add energy to the fluid directly by hydraulic force. This causes a pressure increase in the fluid, which builds up until the discharge valve opens and the fluid is displaced into the system. As a result, the head-capacity curve of a reciprocating pump at constant speed is essentially a straight vertical line. A positive displacement pump will produce fluid flow that corresponds to its particular speed, regardless of the pressure it is pumping against. Fluctuations in pressure will not change the flow. The pump capacity changes with the pump speed, but its pressure capability remains unchanged. The pump can develop any pressure at any operating speed, within its mechanical limitations.

Since reciprocating pumps can develop pressure in excess of their design rating, it is necessary to protect the pump and piping with a pressure relief valve in the discharge line. Such a valve will open at a given pressure, limiting the pressure in the system by dumping the flow either to waste or to the suction tank.

Reciprocating pumps can handle a wide variety of liquids—including those with extremely high viscosities, high temperatures, and high slurry concentrations—as a result of the pump's basic operating principle. That is, it adds energy to the fluid by direct application of force, rather than by acceleration. In the case of a highly viscous fluid, for instance, it is only necessary to be sure that the fluid gets into the pumping chamber because what enters must, eventually, be displaced out. At times it may be necessary to slow the pump down so that the viscous fluid has time to fill the pumping chamber on each stroke. The head on the viscous fluid must be sufficient to move the fluid into the pump cylinder.

A special case of a positive displacement pump is the blow case. This configuration consists of two pressure chambers which alternately are filled with liquid. When a chamber is filled, air or steam is forced into the chamber, causing the contents to be discharged into the system. The two chambers alternate in this action, so there is a fairly constant discharge. Blow case pumps are popular for pumping hot condensate, since there is no heat loss and flashing liquid can be transferred. See Fig. 2-4.

Rotary Pumps: These rotary devices constitute a large class, used for relatively low flow and moderate pressures. Certain configurations,

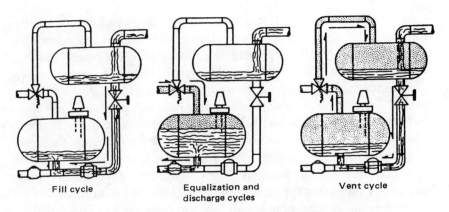

Fill cycle   Equalization and   Vent cycle
              discharge cycles

Fig. 2-4. Operation of Blow Case Pump

however, especially when used in hydraulic systems, may develop several thousand pounds per square inch. The chart, Fig. 2-5, illustrates the many configurations possible. Fig. 2-9, Items 1 through 16, illustrates the variety of arrangements possible.

Flexible vane pumps are largely used for low pressure water pumping; gear pumps are used for oil and similar fluids which have some lubricating value. The two- and three-screw pumps have wide use in pumping fuel oil of various viscosities, and possibly hot. The single-screw pump is arranged with a twisted spiral rotor turning in a relatively soft rubber stator.

Multiple screw: The multiple screw pump, with two or three intermeshing long screws, has been made possible because of the improvement of manufacturing techniques during the current century. Multiple screw pumps are used to handle a variety of fluids, including greases, rubber, lube oils and fuel oil. Where suction lift, high pressure, high capacity, or a broad range of viscosity and temperatures are required, screw pumps are a most efficient device. Table 2-1 illustrates the general design limitation of conventional screw pumps.

Figs. 2-6 and 2-7 illustrate two versions of a twin screw pump, utilizing external or internal bearings. The intermeshing screws do not touch; there is a very fine clearance between them. The screws are turned by timing gears, which therefore bear most of the pumping

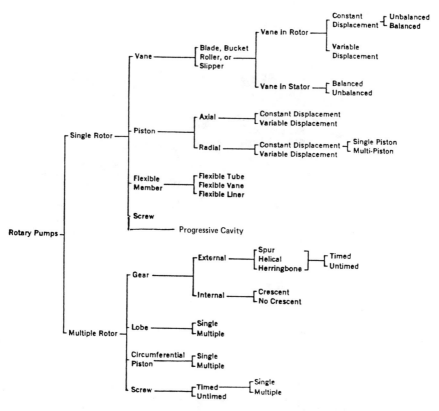

Fig. 2-5. Classes of Rotary Pumps

Table 2-1. General Design Limitations of a Conventional Twin Screw Pump

	Internal Design	External Design
maximum capacity (gpm)	4000	10,000
maximum differential pressure (psi)	2000	2500
viscosity (ssu)	$150-5 \times 10^4$	$32-200 \times 10^4$
temperature (°F)	325	850
speed range (rpm)	10–2750	10–1750
npshr (ft of fluid)	3 ft–100 ft	3 ft–100 ft

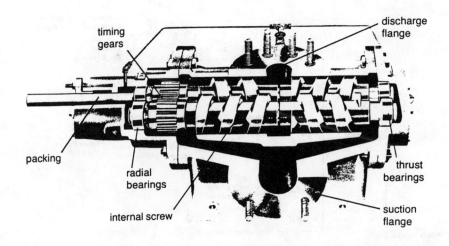

Fig. 2-6. Internal Bearing Twin Screw Pump Design
(From POWER & FLUIDS, Vol. 7, No. 1, 1981, Worthington Group • McGraw Edison Co.)

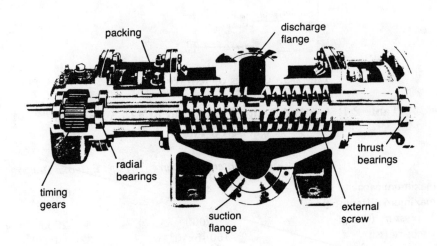

Fig. 2-7. External Bearing Twin Screw Pump Design
(From POWER & FLUIDS, Vol. 7, No. 1, 1981, Worthington Group • McGraw Edison Co.)

load. When the internal gear design is used, bearings and timing gears are located in the pumping chamber, and are lubricated by the fluid being pumped. An external design pump has the bearings and gears located in a separate lube oil chamber outside the pumping cavity. With the broad limits shown in Table 2-1, an internal design is recommended if the fluid being pumped will lubricate the anti-friction bearings. Lubricating fluids are usually characterized by viscosities above 150 ssu and free of hard abrasive and corrosive compounds. An external design is recommended when fluid lubricating characteristics are questionable.

In the pump design, the fluid enters at the outer ends of the screws, and is moved inward to the center discharge port. The seals are therefore not subjected to high discharge pressures. The screw element must be designed to withstand the hydraulic pressure created within the screw channels.

Axial force is balanced by using right- and left-hand screws on the same shaft; the radial force, however, causes shaft deflection, which is resisted by the fluid film trapped between the screws and the pump body. An excellent discussion is contained in Worthington Company's "Power & Fluids," for 1981, Vol. 7, No. 1.

Progressive cavity: This concept is an extremely simple and rugged, positive displacement pump of unique design. It consists of a single helical rotor turning inside a resilient stator which has an opposite spiraling double helix cavity that is constantly opened and resealed with each revolution of the rotor. Delivery is essentially pulseless. Since delivery is proportional to speed, it makes an excellent metering pump. The rate of discharge is only slightly affected by discharge pressure. Pumping is gentle, so that shear-sensitive materials may be handled. Polymers, highly viscous materials, and high solid content sludges may be readily moved. Added capacity for pumping very low viscosity and gaseous fluids is achieved by the use of an elastomeric stator that results in an interference fit between the rotor and stator along the seal line. These pumps are self-priming. They must be driven at speeds much less than occasioned by direct electrical drive. Since torque to the rotor is delivered by a Cardan (Universal) joint, it is important to the life of the pump to have adequate capacity for relative motion between the orbiting rotor and the fixed drive shaft. Sealing is not a problem, as the seals at the discharge end are not required. Fig. 2-8 illustrates the construction.

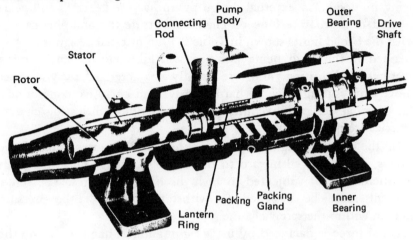

Fig. 2-8. Progressive Cavity Pump
(From ProCav catalog, Peabody Barnes, Mansfield, Ohio)

The several pumps not noted are used wherever their specific configuration best suits the application. For instance, flexible tube pumps are very simple, cheap and easy to repair. They will not produce significant pressure, but they can move fluid samples or chemicals in laboratory applications. The Standards of the Heat Exchange Institute contain simple descriptions of the operating mechanisms of the many rotary pumps shown in Fig. 2-9.

Basic types: There are four basic types in the single rotor pump class and also four basic types in the multiple rotor pump class.

Vane: In this type, the vane or vanes, which may be in the form of blades, buckets, rollers, or slippers, cooperate with a cam to draw fluid into and force it from the pump chamber. These pumps may be made with vanes in either the rotor or stator and with radial hydraulic forces on the rotor balanced or unbalanced. The vane-in-rotor pumps may be made with constant or variable displacement pumping elements. Item 1 of Fig. 2-9 illustrates a vane-in-rotor constant displacement unbalanced pump. Item 2 illustrates a vane-in-stator constant displacement unbalanced pump.

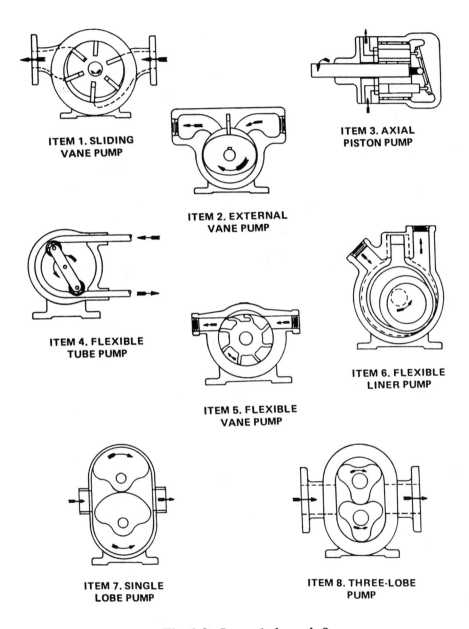

ITEM 1. SLIDING
VANE PUMP

ITEM 2. EXTERNAL
VANE PUMP

ITEM 3. AXIAL
PISTON PUMP

ITEM 4. FLEXIBLE
TUBE PUMP

ITEM 5. FLEXIBLE
VANE PUMP

ITEM 6. FLEXIBLE
LINER PUMP

ITEM 7. SINGLE
LOBE PUMP

ITEM 8. THREE-LOBE
PUMP

Fig. 2-9. Items 1 through 8

(By permission of the Hydraulic Institute)

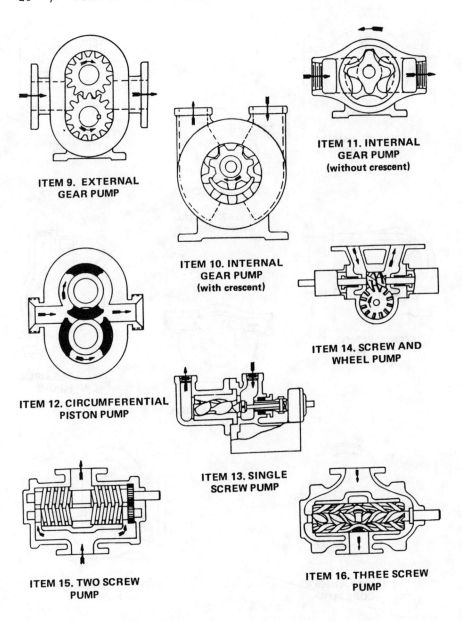

ITEM 9. EXTERNAL
GEAR PUMP

ITEM 10. INTERNAL
GEAR PUMP
(with crescent)

ITEM 11. INTERNAL
GEAR PUMP
(without crescent)

ITEM 12. CIRCUMFERENTIAL
PISTON PUMP

ITEM 13. SINGLE
SCREW PUMP

ITEM 14. SCREW AND
WHEEL PUMP

ITEM 15. TWO SCREW
PUMP

ITEM 16. THREE SCREW
PUMP

Fig. 2-9. Items 9 through 16
(By permission of the Hydraulic Institute)

Piston: In this type, fluid is drawn in and forced out by pistons which reciprocate within cylinders with the valving accomplished by rotation of the pistons and cylinders relative to the ports. The cylinders may be axially or radially disposed and arranged for either constant or variable displacement pumping. All types are made with multiple pistons except that the constant displacement radial type may be either single or multiple piston. Item 3 illustrates an axial, constant displacement piston pump.

Flexible member: In this type, the fluid pumping and sealing action depends on the elasticity of the flexible member(s). The flexible member may be a tube, a vane, or a liner. These types are illustrated in Items 4, 5 and 6 respectively.

Lobes: In this type, fluid is carried between rotor lobe surfaces from the inlet to the outlet. The rotor surfaces cooperate to provide continuous sealing. The rotors must be timed by separate means. Each rotor has one or more lobes. Items 7 and 8 illustrate a single- and a three-lobe pump, respectively.

Gear: In this type, fluid is carried between gear teeth and displaced when they mesh. The surface of the rotors cooperate to provide continuous sealing and either rotor is capable of driving the other.

External gear pumps have all gear rotors cut externally. These may have spur, helical, or herringbone gear teeth and may use timing gears. Internal gear pumps have one rotor with externally cut gear teeth meshing with an externally cut gear. Pumps of this class are made with or without a crescent-shaped partition. Item 9 illustrates an external spur gear pump. Items 10 and 11 illustrate internal gear pumps with and without the crescent-shaped partition.

Circumferential piston: In this type, fluid is carried from inlet to outlet in spaces between piston surfaces. There are no sealing contacts between rotor surfaces. In the external circumferential piston pump, the rotors must be timed by separate means, and each rotor may have one or more piston elements. In the internal circumferential piston pump, timing is not required, and each rotor must have two or more piston elements. Item 12 illustrates an external multiple piston type. The shadowed portion of the figure represents the rotating parts.

Screw (single): In one type, fluid is carried between rotor screw threads and is displaced axially, as they mesh with internal threads on the stator. The rotor threads are eccentric to the axis of rotation. This type is illustrated in Item 13 of Fig. 2-9. Another type of single screw pump is shown in Item 14. This type depends upon a plate wheel to seal the screw so that there is no continuous cavity between the suction and discharge.

Screw (multiple): In this type, fluid is carried between rotor screw threads and is displaced axially as they mesh. Such pumps may be timed or untimed. Item 15 illustrates a timed screw pump. Item 16 illustrates an untimed screw pump with three rotors.

DYNAMIC (KINETIC)

Centrifugal pumps: All centrifugal pumps utilize but one pumping principle: the impeller rotates the fluid at high velocity, building up a velocity head. At the periphery of the pump casing, the fluid is directed into a volute (sometimes called a diffuser). The volute most often has a constantly increasing cross-sectional area along its length, so that as the fluid proceeds along the channel, its velocity is reduced. Because the energy level of the fluid cannot be dissipated at this point, the conservation of the energy level requires that when the fluid loses velocity energy as it moves along the channel, it must increase the energy related to pressure. That is, the pressure of the fluid increases. Although this principle is valid for most pumps in this class, there are some instances in which the dynamics are somewhat modified.

Fig. 2-10 formats the categories into which this class may be divided. The pure centrifugal radial, the mixed, and axial flow configurations are better described in the following paragraphs.

Centrifugal pumps may be single stage, having a single impeller, or they may be multiple stage, having several impellers through which the fluid flows in series. Each impeller in the series increases the pressure of the fluid at the pump discharge. Pumps may have thirty or more stages in extreme cases. In the case of centrifugal pumps, a correlation of pump capacity, head, and speed at optimum efficiency is used to classify the pump impellers with respect to their specific geometry. This correlation is called Specific Speed, and is an important parameter for analyzing pump performance.

CENTRIFUGAL PUMP CLASSES

Radial flow: A pump in which the pressure is developed principally by the action of centrifugal force. Pumps in this class, with single inlet impellers, usually have a specific speed below 4200, and with double section impellers, a specific speed of below 6000. In pumps of this class the liquid normally enters the impeller at the hub and flows radially to the periphery. (Fig. 2-10A)

Mixed flow: A pump in which the head is developed partly by centrifugal force and partly by the lift of the vanes on the liquid. This type of pump has a single inlet impeller with the flow entering axially and discharging in an axial and radial direction. Pumps of this type usually have a specific speed from 4200 to 9000. (Fig. 2-10 B)

Axial flow: A pump of this type, sometimes called a propeller pump, develops most of its

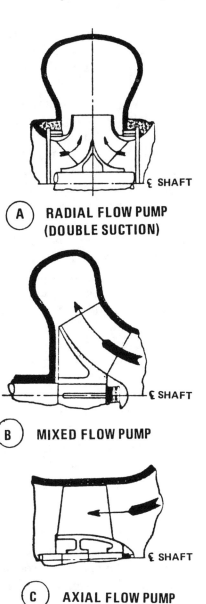

A **RADIAL FLOW PUMP (DOUBLE SUCTION)**

B **MIXED FLOW PUMP**

C **AXIAL FLOW PUMP**

Fig. 2-10
(Permission of the Hydraulic Institute)

head by the propelling or lifting action of the vanes on the liquid. It has a single inlet impeller with the flow entering axially and discharging nearly axially. Pumps of this type have a specific speed above 9000. (Fig. 2-10C)

Turbine pumps (regenerative): The regenerative turbine pump offers many advantages in the area of low flow and moderate-to-high pressure. The turbine pump is reasonably efficient at low flow. Pulsations are less than is the case for a centrifugal pump. Turbine pumps are very versatile because of their compactness and single stage pressure capacity. They are recognized for their usefulness in low-flow, high-head applications. The turbine pump offers higher heads than a centrifugal pump by a factor of five or more compared to an equivalent number of centrifugal pump stages. The head capacity curve is steep, so the pump has some of the characteristics of a positive displacement pump. The steep curve allows a good degree of flexibility in application, but it also requires a relief valve for safe operation. In addition, if operated past the design point, the motor may overload. The efficiency curve tends to fall off sharply on either side of the best efficiency point.

Turbine pumps having a top center line discharge, are self-venting, and have the ability to handle vapors without vapor lock. This characteristic allows the pump to handle boiling liquids and liquefied gases at suction heads but slightly over the vapor pressure of the fluid. At very low flows, the pump has better efficiency than a small centrifugal pump. Construction normally utilizes close running clearances. Used only for clean liquid application, they are popular for small boiler feedwater pumping application. Viscous fluids up to 500 ssu can be pumped, but above this value the pump is not applicable.

Turbine pumps derive their name from the many buckets machined into the periphery of the rotating impeller. Heads over 900 feet are readily developed in a two-stage pump. The impeller, having very close axial clearance, and utilizing pump channel rings, has minimum recirculation losses. The channel rings provide a circular channel around the blades of the impeller, from the inlet to the outlet. Liquid entering the channel from the inlet is picked up immediately by the buckets on both sides of the impeller and pumped through the channel by a shearing action. The shearing action and the flow through the impellers is illustrated by Fig. 2-11 A and B. The process is repeated over and over, each cycle imparting more energy until the liquid is discharged. In a two-stage pump, the liquid is directed to a second impeller, where the entire process is repeated, doubling the discharge head.

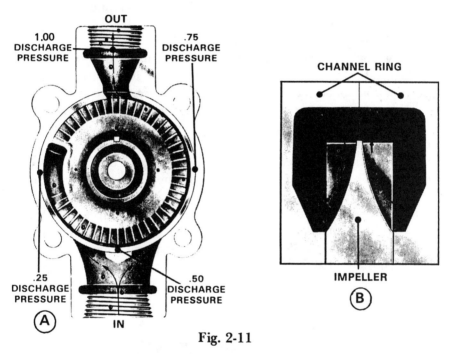

Fig. 2-11

Laboratory observations indicate the pressure head developed is a function of the number of recirculations that take place in the flow from suction to discharge. The special brake horsepower curve with its increased power input at increased head serves to highlight the regenerative character of this special centrifugal design. Maximum shut-off head developed is normally about ten times the shut-off head of a single-stage centrifugal impeller of the same diameter running at the same speed.

The following phenomena lend themselves to conclusions separating this equipment from rotary positive displacement pumps:

1. Liquid can be forced through the pump only with the impeller locked. Only a partial pressure drop will be noted.

2. The impeller can be run for short periods against full shut-off head on the heavier models. This would be destructive of close clearance rotary pumps without relief valves.

3. Turbine pumps cannot be used on highly viscous liquids.

4. Vapor can be moved only when the pump is liquid-filled.

5. Head capacity curve is very steep.

The Pitot pump: The basic concept behind the Pitot pump is remarkably simple: liquid enters the intake manifold and passes into a rotating case where centrifugal force accelerates it. A stationary pickup tube positioned on the inner edge of the case, where pressure and velocity are greatest, converts the centrifugal energy into a steady, pulsation-free, high-pressure stream. The simplicity of the pump lies in the fact that there is only one rotating part, the seal is exposed only to suction pressure, and no seal is required at the high-pressure discharge. The pump, turning at speeds from 1325 to 4500 rpm, will generate heads roughly four times that of a single-stage centrifugal pump operating at a similar speed. Single-state heads up to 3000 feet are readily attainable, even in sizes up to 200 gpm. Efficiency is somewhat less than for conventional centrifugal pumps. In small sizes, required NPSH may be as low as 4 feet, but in the large sizes, caution must be used. Required NPSH may be as high as 100 feet. Fig. 2-12 illustrates the Pitot pump.

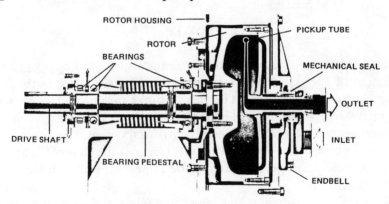

Fig. 2-12. Kobe Roto-Jet Pump

The Liquid ring pump: The liquid ring pump (Fig. 2-13) is an unusual pump in that it is not used for conventional water pumping applications. However, as a vacuum pump—even as a low-pressure gas or air compressor, it has numerous advantages. It has only one moving part, a balanced rotor. All functions of mechanical pistons or vanes are actually performed by a rotating band of liquid compressant. While power to induce rotation is transmitted by the rotor, this ring of liquid follows a path around the cylindrical interior of the pump body. The rotor axis is offset from the body axis. It is a cross between a dynamic and a positive displacement pump. As the figure shows, liquid compressant almost fills, then almost empties, each rotor chamber during a single revolution. That sets up the piston action. Inlet connections and pressure discharge connections are separated by ported openings in the stationary inner cones.

These pumps are ideal for providing condenser vacuums, for suction vacuum on large centrifugal pump systems, and for dewatering applications. When used as a condenser exhauster, the pump may be double-staged in conjunction with an eductor.

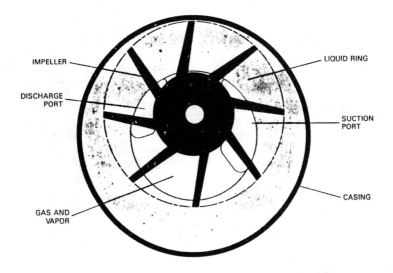

Fig. 2-13. End View — Liquid Ring Pump Chamber
(Source: Ochsner Pump Co.)

Screw centrifugal pump: This is a recent design by the Lawrence Pump Company, incorporating a large diameter screw instead of the more common radial impeller found in centrifugal pumps. Thick sludge and large solids can be moved because of the low NPSH requirements resulting from the use of the inducer-like impeller. As the pumped material enters at a low entrance angle, a low shear, low turbulence condition exists, resulting in very gentle handling of the fluid. The gentle handling makes it possible to pump live fish, and slurries of fruits or vegetables, without undue breakup of the constituents. The pump can also be operated in reverse rotation, which is advantageous for clearing clogged suction lines. Fig. 2-14 shows the configuration.

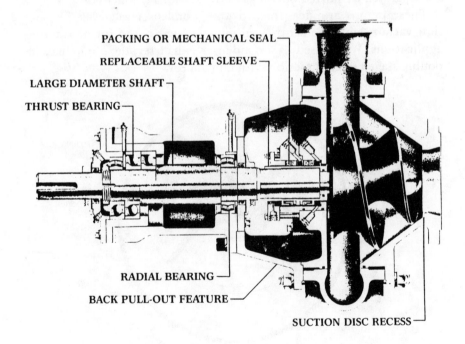

PACKING OR MECHANICAL SEAL
REPLACEABLE SHAFT SLEEVE
LARGE DIAMETER SHAFT
THRUST BEARING
RADIAL BEARING
BACK PULL-OUT FEATURE
SUCTION DISC RECESS

Fig. 2-14. The Screw Centrifugal Pump
(Source: Lawrence Pump Company)

Viscous drag or disc pump: Fig. 2-15. The concept behind the disc pump is over 100 years old. Nicola Tesla, the genius who invented the alternating current motor and generator, invented and patented the first practical working model of the pump in 1913. In spite of the many advantages demonstrated for this concept, the disc pump remained a mere curiosity until the "drag" concept was reexamined and reintroduced in the mid-seventies. A disc, or drag, turbine was developed, but was dropped for technical reasons. The disc pump, however, as developed by Max Gurth, appeared to be a technically feasible, as well as a useful, concept. The theory of operation incorporates two obscure but salient principles of fluid mechanics— boundary layer and viscous drag. These phenomena occur simultaneously whenever a surface is moved through a fluid. The boundary layer phenomenon occurs in the disc pump when fluid molecules "lock" onto or embed on the surface roughness of the disc rotors. A dynamic force field is developed, which produces a strong radially accelerating friction force gradient within and between the molecules of the fluid and the discs, creating a boundary layer effect. The resulting frictional resistance force field between the interacting elements and the natural predilection of fluids to resist separation of its own continuum, creates the "adhesion" phenomenon known as viscous drag. These effects, acting together, are the prime movers in transferring the necessary tangential and centrifugal forces to propel the fluid with increasing momentum toward the discharge outlet at the periphery of the discs.

Advantages claimed for this pump are minimum wear with abrasive materials, gentle handling of delicate fluids, ability to easily handle highly viscous or heavily loaded fluids, and freedom from vapor lock. Disc type pumps are manufactured by Disco Flow as well as by U.S. Pump and Turbine in Paramount, California.

The open screw pump: Fig. 2-16. This pump is discussed as an indication to the reader that there are numbers of pump configurations that do not conform to the more or less classical forms, as we have discussed. The forerunners of this device are to be found in antiquity. The occasional designation of Archimedes' screw takes the design back to the early Greeks. The pump consists of a U-shaped channel, into which a rotating screw fits closely. The channel, angled

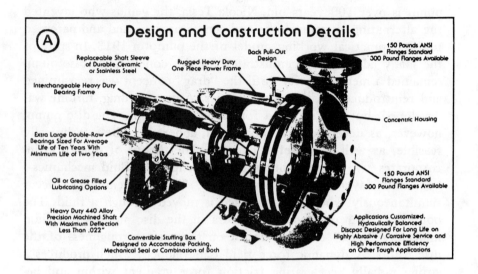

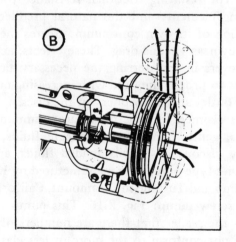

Fig. 2-15. Disc Pump
(Source: DISCFLO Disc Pumps, Santee, CA)

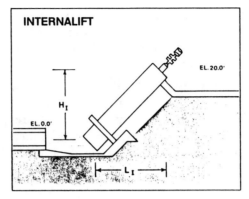

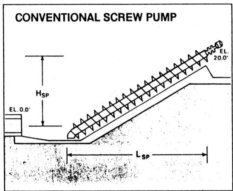

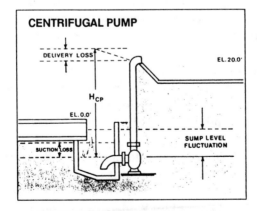

Fig. 2-16. Open Screw Pump
(Source: CPC Engineering Corporation)

at inclinations up to forty-five degrees, takes water from a lower level and literally "screws" the water from the lower to a higher level. The screw, of course, does not develop any pressure, as it is merely a conveyor. Modern forms of this pump are quite large. Used extensively in wastewater plants for moving contaminated water, they are excellent for this purpose, as there is very little to go wrong. The large sizes, with a closely-fitted screw, are quite efficient. One version surrounds the screw within a large tube, and the whole assembly is then rotated. All bearings are thus outside of the liquid and there is no leakage.

SOLIDS HANDLING PUMPS

A popular pump for handling extremely dirty fluids, slurries, and abrasive materials is the air-powered, double-diaphragm pump—a unique device with many advantages. Safety, simplicity, versatility and economy are features of this pump.

There are two diaphragm chambers and two flexible diaphragms. Clamped in sandwich fashion at their outer edges, the diaphragms are connected by a shaft and move simultaneously in a parallel path. Compressed air directed into the left chamber moves both diaphragms to the left and exhausts air from the right chamber.

On completion of a stroke, an air distribution valve automatically transfers the air flow to the right chamber, reverses the diaphragm movement and exhausts air from the left chamber.

This continuous, reciprocating motion thus creates an alternate intake and discharge of the liquid being pumped into and out of each chamber. As a portion of the liquid is drawn into the right side, an equal portion is simultaneously discharged from the left, and vice versa.

Check valves—two on each side—automatically control the flow into and through the chambers and out the discharge.

Fig. 2-17 illustrates the operation of this pump. The illustration is that of a pump manufactured by Wilden, one of several manufacturers of pumps using this principle of pumping.

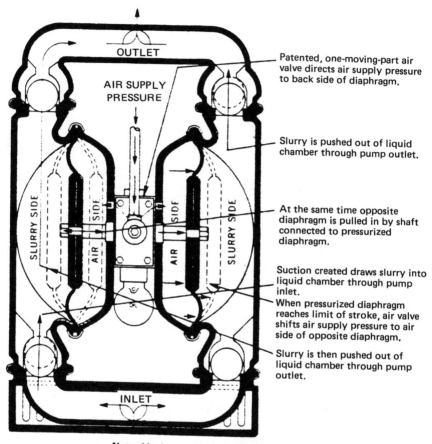

OUTLET

AIR SUPPLY PRESSURE

SLURRY SIDE

AIR SIDE

AIR SIDE

SLURRY SIDE

INLET

Patented, one-moving-part air valve directs air supply pressure to back side of diaphragm.

Slurry is pushed out of liquid chamber through pump outlet.

At the same time opposite diaphragm is pulled in by shaft connected to pressurized diaphragm.

Suction created draws slurry into liquid chamber through pump inlet. When pressurized diaphragm reaches limit of stroke, air valve shifts air supply pressure to air side of opposite diaphragm.

Slurry is then pushed out of liquid chamber through pump outlet.

Note: Maximum non-compressible solid the pump can pass is limited by clearance between ball valve and valve seat: M-2 1/8", M-4 3/16", M-8 1/4", M-15 1/2"

Compressed air is applied directly to the liquid column separated by elastomer diaphragms. This balanced load removes the mechanical stress from the diaphragms to allow high heads and thousands of hours of diaphragm life.

The pumping volume is controlled by easy inlet air adjustments, from a few gallons per hour to over fourteen thousand gallons per hour with the same unit. A by-pass valve is not required because the pump stops when discharge pressure equals air inlet pressure. The pump can also run dry indefinitely without damage. Our double diaphragm design cuts velocity through the pump to half total discharge velocity. The most abrasive slurries produce little wear effect on pump parts.

Fig. 2-17
(Source: Wilden Pump)

3

Specialized Pumps

In the previous chapter, we described how and why the various pumping principles are practically used. We should now have some acquaintance with some of the specialized pump configurations which have various, diverse requirements.

CANNED

There are many situations in which any pump leakage is absolutely prohibited. A good example is pumps which move nuclear coolants, or slurries. The most effective way to inhibit leakage absolutely is to seal the pump, or the pump and motor, in a liquid-tight container. Such a unit is said to be "canned." A canned-motor pump is actually a centrifugal pump and a squirrel-cage induction electric motor built into a common, hermetically-sealed container.

Fig. 3-1A illustrates such a unit. The one moving part is an integrated impeller-rotor assembly, driven by the rotating field of the motor. A nonmagnetic corrosion-resistant liner, or "can," surrounds the rotor-impeller assembly, so that the pumped fluid does not wet the stator windings. The assembly bearings are lubricated by the pumped fluid which is circulated through a filtered bypass system.

The stator may be cooled by the pumped fluid, or the stator of the motor may be open to the atmosphere, and is thus air cooled. Another type contains both motor and pump sealed in separate compartments, both hermetically sealed. The advantage of this latter arrangement is containment of the fluid in the event of a rupture of the stator liner. All canned pumps have power losses because of large stator-to-rotor gap in the sealed motor type. Moreover, the choice of bearings is critical for pump reliability in any sealed pump. There is, fortunately, a large choice of bearing and shaft materials, which should be selected with due regard for temperature, corrosion resistance, and compatibility with the pumped fluid as lubrication.

MAGNETIC DRIVE

A second type of sealless pump is one in which only the pump itself is totally enclosed in a sealed container. The motor is external to the can, and the pump is driven by matched, opposed magnets, one on the motor and the other on the pump. The magnetic drag of the motor magnets acts through the sealed nonmagnetic container wall, thus driving the magnets in the can, and hence, the pump. The driving and driven magnets may be face-to-face, or the driving magnet may encircle a cup containing the driven magnet. The two magnets align themselves pole-to-pole and rotate together with no slippage until the decoupling limit is exceeded. Decoupling occurs when the pump load exceeds the coupling torque between the magnets. This feature can act as a safety device to protect pump and motor from inadvertent damage. To recouple the magnets, bring the motor to a complete stop, eliminate the cause of decoupling, and restart.

In magnet drive pumps (Fig. 3-1B) the driven magnet is exposed to the fluid being pumped. To insure that the fluid remains contaminant-free, the magnet must be encapsulated in an inert coating, or in stainless steel.

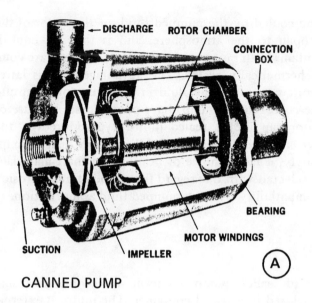

CANNED PUMP

DISCHARGE — ROTOR CHAMBER

CONNECTION BOX

BEARING

MOTOR WINDINGS

SUCTION

IMPELLER

(A)

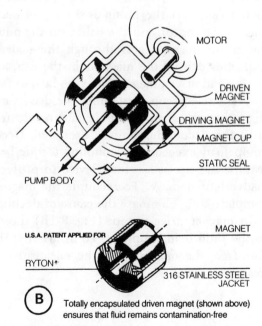

MAGNETIC DRIVE PUMP

MOTOR

DRIVEN MAGNET

DRIVING MAGNET

MAGNET CUP

STATIC SEAL

PUMP BODY

U.S.A. PATENT APPLIED FOR

RYTON*

MAGNET

316 STAINLESS STEEL JACKET

(B) Totally encapsulated driven magnet (shown above) ensures that fluid remains contamination-free

Fig. 3-1

HIGH TEMPERATURE

Pumps are frequently faced with the necessity of handling high temperature fluids, varying from boiling water to liquid metals. The elements sensitive to temperature are the wetted components, and the seals. A gamut of techniques is utilized, starting with high temperature materials and seals, to cooled seals and cooled wetted parts. Perhaps the ultimate in high temperature design is the liquid metal pump, which utilizes high temperature materials as well as forced cooling of critical components. An intermediate temperature pump, the KSB Etanorm (Fig. 3-2) utilizes a self-cooling design to permit operation at temperatures up to 600°F, with heat transfer oils.

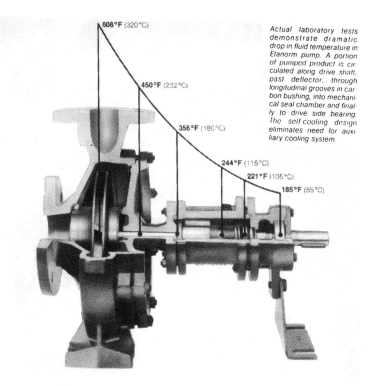

Actual laboratory tests demonstrate dramatic drop in fluid temperature in Etanorm pump. A portion of pumped product is circulated along drive shaft, past deflector, through longitudinal grooves in carbon bushing, into mechanical seal chamber and finally to drive side bearing. The self cooling design eliminates need for auxiliary cooling system.

Fig. 3-2. Intermediate Temperature Pump
(Source: KSB Inc., Hauppauge, NY)

TURBINE DRIVE

A combination of steam turbine driving a centrifugal pump is an ideal arrangement for high pressure pumps. Aside from the benefits of variable speed, several important benefits are apparent: high rotor speed can be utilized, resulting in high discharge pressures from a single stage; if turbine and pump impellers are mounted back-to-back, high-pressure seals on the pump are simplified; alignment problems are eliminated; the arrangement can be made extremely compact; a single shaft serves both turbine and pump, eliminating a coupling and several bearings. Fig. 3-3 shows an FMC unit.

Turbo Pumps are single-or double-stage centrifugal diffuser-type pumps driven by single stage, one or two row, impulse-type steam turbines. The pump and turbine are mounted at either end of a short, rigid shaft. The rotating assembly is supported by a set of high-speed, precision ball bearings, with the pump and turbine overhung over these bearings.

Fig. 3-3. Cutaway of Turbo Pump

(Source: FMC Corporation, Englewood, NJ)

OVERHUNG OR CANTILEVER IMPELLER

In some cases, it is considered undesirable to subject bearings or packing to the pumped fluid. Temperature, corrosive, or erosive conditions may not favor a conventional configuration. In such cases, an overhung impeller is used. In this design, the pump is mounted vertically, and the impeller is mounted on a heavy shaft suspended, without bearings, in the casing. The shaft is supported by bearings at some distance from the impeller, and completely isolated from the pumped fluid. This arrangement may be termed exotic. An extremely heavy shaft is required, sufficiently rigid so that any vibration of the impeller does not cause significant shaft deflection. The bearings, at some distance from the impeller, must be designed to accept magnified radial loads due to any imbalance at the impeller. The consequence of the design, however, is a reliable pump free of many of the problems encountered by competitive conventional pumps.

A good example of a cantilevered pump, manufactured by Lawrence Pumps Inc., is shown in Fig. 3-4. Various casing and impeller styles available for the Lawrence pump are shown in Figs. 3-5 through 3-12. Although the illustrations apply to the Lawrence pumps, they are illustrative of the diversity of liquid ends applicable to centrifugal pumps for different classes of service.

TRASH PUMPS

Trash in waste water plants presents a serious problem in pumping, the more so if it is stringy or consists of large pieces, that is, measuring up to several inches in size. Size itself is not an insurmountable problem, as many impellers are designed to pass relatively large balls. When combined with sticky or stringy materials, however, complete stoppage of flow may be anticipated unless the pump is of special design. One method of moving difficult fluids is by using an impeller which is completely recessed into the pump housing. Liquid is pumped by the rotation of the fluid in the housing, induced by the movement of the recessed impeller. Fluid does not flow through the impeller at all; thus, anything that comes into the intake part will be ejected through the discharge.

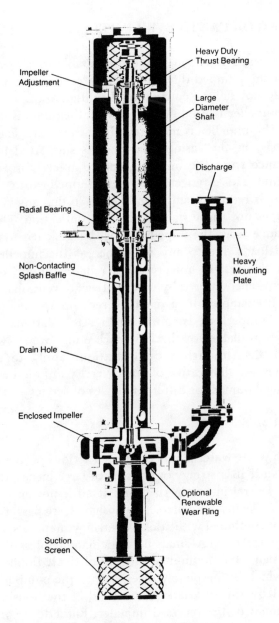

Fig. 3-4. Cantilevered Pump

(Source: Lawrence Pumps, Inc., Lawrence MA)

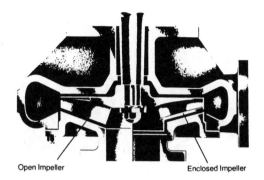

Open Impeller   Enclosed Impeller

This pump is specifically designed to handle clean liquids. It is commonly used for chemicals and water solutions, as well as molten salts, molten metals, mercury, and explosive liquids. Open impellers are used for most services, including viscous liquids. Enclosed impellers are used on high head applications. Sealing rings contain recirculation. Replaceabel wear rings are optional.

Fig. 3-5. Specialized Pump for Handling Clean Liquids

(Source: Lawrence Pumps, Inc., Lawrence, MA)

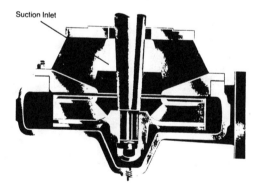

Suction Inlet

This design is used primarily for handling entrained air, gaseous liquids, high vapor pressure liquids, flotation, and skimming. Pumps with open, enclosed, or slurry type impellers are used as required.

Fig. 3-6. Top Suction Pump

(Source: Lawrence Pumps, Inc., Lawrence, MA)

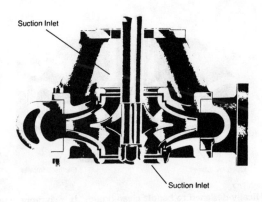

Double suction pumps are used for high flow, high head services. They require less NPSH than single pumps. Double suction balances axial thrust and double volute balances radial forces.

Fig. 3-7. Double Suction Pump

(Source: Lawrence Pumps, Inc., Lawrence, MA)

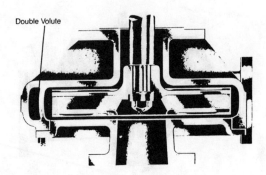

The double volute design balances radial forces within the pump casing and combines the flow into one outlet. This provides a more efficient operation than the older dual discharge design. These pumps are available in clean liquid and light slurry designs.

Fig. 3-8. Double Volute Pump

(Source: Lawrence Pumps, Inc., Lawrence, MA)

Optional Suction Disc Liner

Slurries require specially designed pumps. The impellers and casings have large passageways and liberal clearances. Impeller repelling vanes are added to reduce wear and recirculation. Heavy wall thickness on the wearing parts provide for longevity. Suction and hub disc liners are optional (suction disc liner shown). Slow RPMs are critical since pump wear varies between the square and cube of the RPM ratio.

Fig. 3-9. Pump for Slurries

(Source: Lawrence Pumps, Inc., Lawrence, MA)

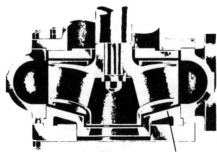

Optional Suction Disc Liner

This casing and impeller configuration differs from other designs due to its large solids handling capability. Single, Double, and Triple impeller vane designs are available. The non-clogging design offers a higher efficiency than the recessed impeller or torque-flow pump and is, therefore, more energy efficient.

Fig. 3-10. Non-Clogging Pump

(Source: Lawrence Pumps, Inc., Lawrence, MA)

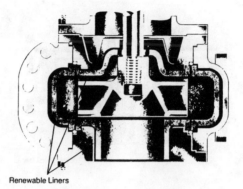

Renewable Liners

Replaceable casing liners are made from various hard metal alloys or Refrax®. The Refrax® lined slurry pump has a replaceable silicon carbide casing liner, and a solid silicon carbide impeller, both of which are next to diamond in hardness. Refrax® is chemically inert to most liquids and is used in highly abrasive applications.

Fig. 3-11. Replaceable Casing Liner
(Source: Lawrence Pumps, Inc., Lawrence, MA)

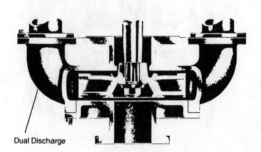

Dual Discharge

The dual discharge pump is the forerunner of the modern double volute design. However, it is less efficient than the double volute pump. The two separate discharge outlets balance the radial forces.

Fig. 3-12. Dual Discharge Pump
(Source: Lawrence Pumps, Inc., Lawrence, MA)

Impellers which are not recessed may be shaped without the usual radial vanes. Single or double vanes, or screws may be used. Fig. 3-13 shows an impeller used in a typical trash pump. Fig. 3-14 shows a pump designed with a recessed impeller.

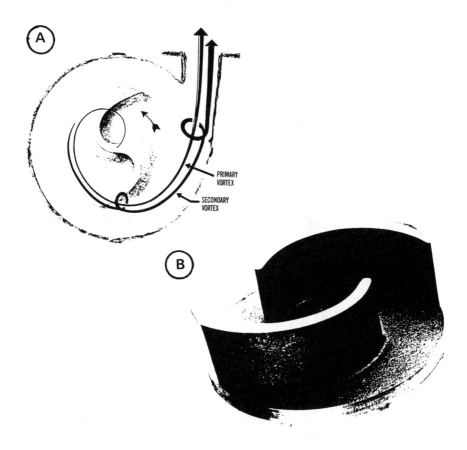

Two distinct vortices are created within the pump to weave and pass any solids through the pump. Trailing edges of the impeller are designed to eliminate low pressure areas. Solids are washed smoothly over the rounded vane edges, down the slope of the impeller, and into the pump discharge. Absence of sharp corners and edges on the impeller prevent "hair-pinning" or hangup of stringy materials.

Fig. 3-13. Double Vortex Pumping Action
(Source: Cornell Pump Co., Portland, OR)

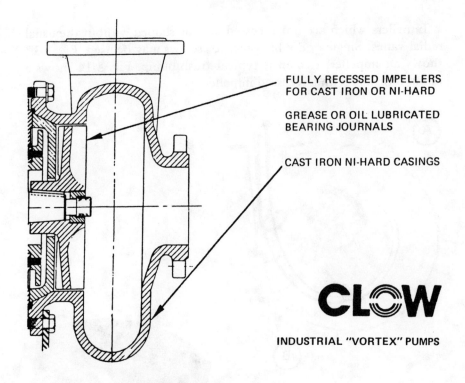

FULLY RECESSED IMPELLERS
FOR CAST IRON OR NI-HARD

GREASE OR OIL LUBRICATED
BEARING JOURNALS

CAST IRON NI-HARD CASINGS

CLOW

INDUSTRIAL "VORTEX" PUMPS

Fig. 3-14. Pump with Recessed Impeller

VERTICAL TURBINE PUMP

The vertical turbine pump was first designed for deep well service and its success in this application is an old story. However, certain inherent performance characteristics and mechanical advantages were recognized and in time the vertical turbine pump was applied to many industrial uses. In the process of widening the scope of application, improvement in the design of this type has developed a pump of such differing characteristics that pump engineers now consider the vertical turbine pump a type unto itself. A more exhaustive discussion is contained in a later chapter. Fig. 3-15 illustrates two typical vertical turbine pumps.

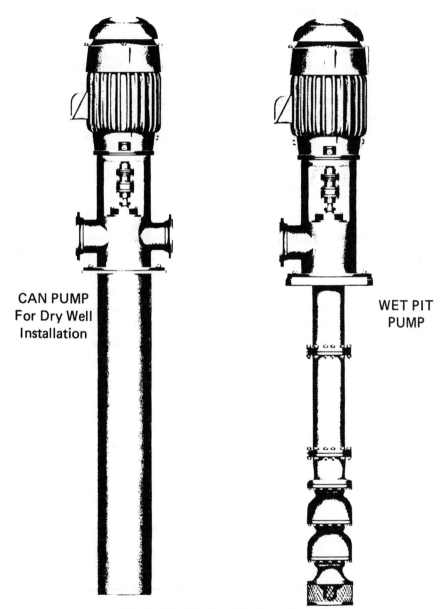

CAN PUMP
For Dry Well
Installation

WET PIT
PUMP

Fig. 3-15. Vertical Turbine Pumps
(Source: Layne & Bowler Inc., Memphis TN)

SUPER-SPEED PUMP

A new design of pump, becoming increasingly popular, is the super-speed pump. Rotating at three to about six times the speed of the driver, exceedingly high heads can be developed with a single stage. In addition, the pump proper is exceedingly small, although the compact size is offset by the addition of the necessary speed-increasing gear box. Other advantages are discussed in a subsequent section.

MATERIALS

Special purpose pumps are now manufactured entirely of plastic, of exotic metals, or any combination, so that almost any chemical can be readily handled. For erosive service, pumps may be lined with ceramic, natural rubber, urethane, or other erosion-resistant materials. Linings to resist heavy erosion are designed to be removed and replaced, as a routine maintenance procedure. Organic materials in pumps are usually temperature-limited. When erosion or corrosion is combined with high temperature, metal wear plates may be used, as well as extremely hard alloys.

4

Pump Arrangements

Pump arrangements fall into several categories, depending upon the feature which is being considered as the basis for classification. The most common categories are:

NUMBER OF STAGES

Single stage, in which the total head is developed by one impeller.

Multi-stage, in which the total head is developed by two or more impellers, in one casing, acting in series.

CASING TYPE

Centrifugal pumps are broadly divided into two classes: volute and diffuser. In the volute pump, the impeller discharges into a single casing channel of a spiral type, so proportionated as to produce an equal velocity of flow at all sections around the circumference and also to gradually reduce the velocity of the liquid as it flows from the impeller to the discharge pipe. In a diffuser pump the impeller discharges into a channel provided with vanes or diffuser. These vanes provide gradually enlarging passages, which function to reduce the velocity of the liquid leaving the impeller and thus efficiently transform velocity head into pressure head.

The diffusion vane casing was introduced into pump design from water turbine practice, where diffusion vanes are indispensable. Early pumps equipped with diffusion vanes were known as turbine pumps.

Fig. 4-1 shows the volute and diffuser arrangements. A third type (Fig. 4-2), in which casing and impeller are concentric, is claimed to eliminate cut water problems inherent in conventional pumps (right). The manufacturer states that this feature reduces inertia loss, power loss, and turbulence and vibration, and thus improves performance, efficiency, and service life.

VOLUTE   DIFFUSER

Fig. 4-1. Schematic of Typical Volute and Diffuser Designs

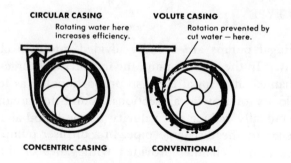

Fig. 4-2. Design with Casing and Impeller Concentric
(Source: DEMCO Pump Co.)

Note that in the volute pump (Fig. 4-1), pressure against the impeller is unbalanced, resulting in an unbalanced load, which is taken by the bearings supporting the impeller shaft. The pump is designed for a radial load on the bearings, and as long as the pump performs at conditions not too far from the design point, the radial loading is accommodated. However, if the pump is operated at less than 30 percent or more than 120 percent of design capacity, the radial load increases drastically, causing early failure of the bearings. More importantly, the unbalanced radial load can cause excessive shaft deflection in areas of fine running clearances, and eccentric loading of mechanical seals, resulting in leakage.

To lower this unbalanced load, double volute diffuser casings are used (Fig. 4-3). In double volute casings, while the pressures are not uniform at part capacity operation, the resultant forces for each 180° volute section oppose and balance each other.

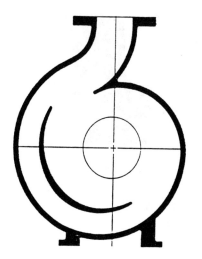

How the Dual Volute Works

All pump volutes are designed to generate uniform radial thrust on the impeller shaft and bearings when operating at the best efficiency point on the pump curve. As a result, there is a minimum of radial thrust on the pump components.

However, when the pump is not operating at the best efficiency point, the casing design no longer balances the hydraulic loads and radial thrust increases.

The dual volute incorporates a flow splitter into the casing which directs the liquid into two separate paths through the casing.

The contour of the flow splitter follows the contour of the casing wall 180 degrees opposite. Both are approximately equidistant from the center of the impeller; thus, the radial thrust loads acting on the impeller are balanced and greatly reduced.

Fig. 4-3. Dual Volute Design of a Typical Ingersoll-Rand ANSI Pump

SUCTION, SINGLE AND DOUBLE

The pump, in its simplest form, comprises an impeller rotating in the pump casing. Liquid is led into the center of the impeller and picked up by the impeller vanes. The rotating impeller accelerates the liquid to a high velocity and discharges it by centrifugal force into the casing and thence out the discharge.

In a single suction impeller, the liquid enters the impeller eye on one side only, while a double suction impeller is, in effect, two single suction impellers arranged back-to-back, in a single casing, so that liquid enters the impeller from both sides. The type of impeller used affects the axial thrust on the impeller shaft. Although thrust bearings are designed to withstand axial thrust, inherent balance is preferable. Theoretically, the double suction pump is in axial balance, while the single suction is not. See Fig. 4-4.

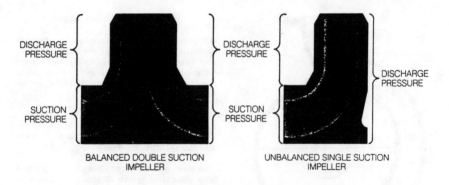

DISCHARGE PRESSURE

DISCHARGE PRESSURE

DISCHARGE PRESSURE

SUCTION PRESSURE

SUCTION PRESSURE

BALANCED DOUBLE SUCTION IMPELLER

UNBALANCED SINGLE SUCTION IMPELLER

Fig. 4-4. Balanced and Unbalanced Axial Thrust in Centrifugal Pumps

However, a single suction impeller with the shaft passing through the impeller eye can be made to balance by using balancing drums, or by reducing stuffing box pressure behind the last stage impeller in multi-stage units, so that the pressure at both ends of the shaft is in approximate balance. The axial balance of an overhung impeller can only be adjusted by adjusting the pressure behind the impeller to counteract the unbalanced axial loads acting on the impeller area equivalent to the shaft or impeller hub sleeve diameter. In cases where a steam turbine impeller and a pump impeller are on the same shaft, it is possible to counteract axial thrust on the pump impeller with the opposite thrust on the turbine impeller. Double suction impellers are favored for large single-stage pumps as being the easiest way to reduce large axial loads, and increase pump capacity and efficiency while maintaining a compact profile.

IMPELLER TYPE

Centrifugal pump impellers are open, closed (or shrouded) or some combination in between. The open impeller is essentially a serial of radial vanes on a backing plate. In the case of trash pumps described earlier, the impeller is open but recessed into the casing. When the radial vanes are shrouded on both sides, the impeller is said to be enclosed. The fluid enters at the open eye of the impeller, and travels to the periphery via an enclosed passageway. Fig. 4-5 shows cross-sections of two typical closed impellers.

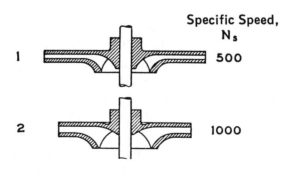

Fig. 4-5

Except in very small pumps, in which it is difficult to cast an impeller with enclosed radial vanes, the most common, or garden variety, of pumps are enclosed impellers. There are some advantages having to do with reduced thrust, good efficiency, reduced tendency towards cavitation. Certain applications require open impellers, however:

1) when fluid containing large solids is to be pumped.

2) fluid containing coarse particles or powder to be mixed. If lumps form they will break up when bounded around in a pump with an open impeller. A closed impeller can get plugged with lumps and become inoperative.

3) the material pumped tends to settle out in the form of a scale or coating on exposed surfaces. When this action takes place in a pump with a closed impeller, the passage gradually decreases in area and pump capacity performance suffers accordingly.

    With this coating action in a pump with an open impeller, the impeller vanes will continually dislodge the scale from the wear plate and thus be at least 50 percent self-cleaning. If they do plug up, the impeller can be easily cleaned.

The influence of impeller configuration is shown in Fig. 4-6, which compares several impeller configurations with their resultant pump curves.

In many cases pump manufacturers have more than one impeller for use with one casing. This enables them to get broader coverage out of fewer parts. It is not unusual for one casing to be used with impellers having a range of diameters.

The impeller with the largest diameter provides the highest head. For reduced heads, the impeller is cut down to match the desired operating point exactly. The shape of the pump curve, important to application, is influenced by the impeller design. Furthermore, one casing may be designed to accommodate small variations in impeller width, thus changing the capacity as well as the shape of the pump curve.

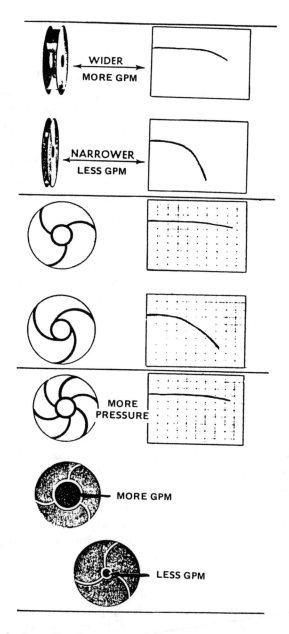

Fig. 4-6. Impeller Configurations with Their Pump Curves

SHAFT POSITION

Pumps are also categorized by the horizontal or vertical position of the shaft. Almost all pumps may be arranged for horizontal or vertical mounting, by varying the supports. Vertical pumps, however, need to have provisions for supporting the weight of the shaft and impeller, which is a downward thrust. A vertically mounted pump should be distinguished from the special category—the vertical turbine pump, noted previously. Vertical pumps have the advantage of saving floor space. Both horizontal and vertical pumps can be mounted in dry pits, below the level of the fluid. Only the suction bell, or part or all of the vertical pump, may be immersed in liquid in a wet pit arrangement.

The wet pit pump has several advantages: it does not require suction piping or suction valve; its suction losses are minimal, resulting in favorable NPSH conditions. Some wet pit pumps are arranged to be completely submerged, including the motor. These are designated as submersible pumps. Fig. 4-7 illustrates a wet sump pump arrangement as used with cooling towers.

MOUNTING AND DISASSEMBLY CONFIGURATIONS

Pumps, whether horizontal or vertical, may be close-coupled on the drive motor, or may be frame-mounted. Frame-mounted pumps are generally larger units, or multiple stage, and can be arranged with intake and discharge in preferred positions. Close-coupled pumps are invariably single end suction, single stage, and are always direct driven, as they are mounted directly on the motor faceplate. They can be arranged for back pull-out, which makes it possible to remove the pump elements without breaking the piping. Frame-mounted pumps are arranged to be horizontally or vertically split. In the horizontal split, the upper half of the casing may be removed for convenient access to the shaft, impellers, bearings and seals. Piping, however, must be disassembled. Frame-mounted pumps may, of course, also be end suction.

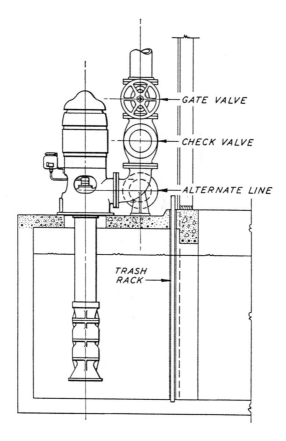

Fig. 4-7. Wet Sump Pump Arrangement

Sliding Mount

An interesting application is the use of a sliding mount for a sub-merged pump. Used where noxious fluids are present, or where maintenance cannot be performed in place, the sliding mount permits the pumps to be lifted to an accessible position for service. The pump can be put in place, raised up and out of the lift station and returned to the discharge connection without personnel having to enter the sump. This is accomplished by sliding the pump on the

T-bar rail which guides it to the discharge connection where it is aligned and joined by the flange coupling. Even after years of exposure in a sump, the T-bar will permit the pump to be raised and lowered without binding. Fig. 4-8 illustrates the method of removal.

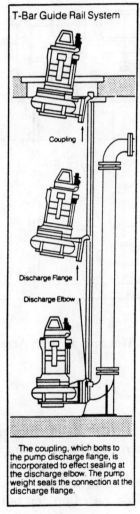

T-Bar Guide Rail System

Coupling

Discharge Flange

Discharge Elbow

The coupling, which bolts to the pump discharge flange, is incorporated to effect sealing at the discharge elbow. The pump weight seals the connection at the discharge flange.

Fig. 4-8

(Source: Davis/EMU Pump Co.)

Pipe Mounted

Various arrangements for connecting suction and discharge piping are available: end suction, side or top discharge; side suction, top or side discharge. An interesting and useful arrangement is the U-shape, in which suction and discharge flanges are both on top of the pump. To a large extent, disassembly methods are governed by the suction and discharge arrangements. A useful arrangement is the pipe-mounted configuration, in which the entire pump and motor is supported solely by the piping connection.

The pipe, of course, must be braced to take the weight. Pipe-mounted pumps are usually small, and vertically mounted. Their great advantage is their lack of floor space requirements.

Vertical "Barrel" Pump

Another useful arrangement is the so-called "barrel" pump. This is a vertically mounted pump having a large number of stages. The intake and discharge flanges are located at some convenient height above the suction end of the pump. The pumped fluid enters the barrel, traverses downward to the suction inlet, then rises to the discharge port. The advantage of a barrel pump is its minimum floor space requirements, as well as the minimum height, since the barrel may be sunk below the level of the floor. Fig. 4-9.

PUMPS AS HYDRAULIC TURBINES

Considering the current search for energy optimization, it is not surprising that the recovery of energy from high pressure water streams is an attractive prospect. The economic aspects, based on the use of commercial power generation hydraulic turbines, are not attractive, however. Limited production, especially in smaller sizes, plus the cost of complex governor gears, account for costs which cannot be reduced by production economics. A practical solution to this dilemma lies in the application of standard pumps as power

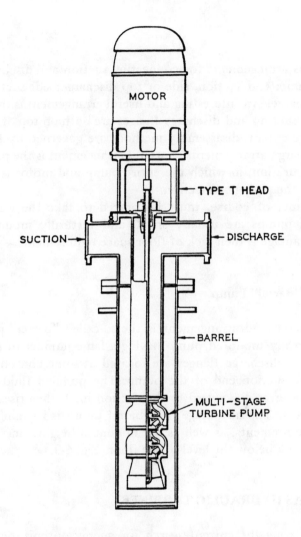

Fig. 4-9. Vertical "Barrel" Pump
(Source: Byron-Jackson Pump Co.)

recovery turbines. Given application where governing equipment is not critical, the large variety of centrifugal pumps introduces new areas in which energy recovery is feasible.

Pumps operating in reverse yield good efficiencies and can be run even more efficiently by operating multiple pumps of various sizes rather than one large conventional hydraulic turbine. Many advantages are apparent: pumps are readily available in many sizes and numerous configurations. They are several generations ahead of commercial turbines in cost effectiveness. As familiar equipment, they are easier to install, operate, and maintain. The many types available—wet pit, dry pit, horizontal, vertical, as examples—make application planning simple. Centrifugal pumps, from radial flow to axial flow geometry, can be successfully operated in reverse. When a centrifugal pump is operated as a normal turbine, flow through the pump is from discharge to suction, and direction of rotation is reversed.

It is also possible to run a pump as a reverse flow turbine, in which case the flow is in the same direction as when pumping (from suction to discharge), and the rotation is also in the normal direction. This condition occurs if one is pumping through an idle pump and results in a relatively low torque output. Whether used as pumps or turbines, a common theory of operation is applicable. Both use the same affinity relationships. This application will be discussed at greater length in a subsequent section.

5

NPSH Controlled Systems

NPSH control or submergence control, as it is commonly called, is inherently automatic and since it requires no control equipment, it has been used very widely in steam power plants to regulate the delivery of condensate pumps. The typical performance curve (Fig. 5-1) for a pump operated on submergence control shows the system-head curve, the NPSH required by the pump and the NPSH available at each of three capacities. At capacities "B" and "C," the available NPSH is exactly equal to the required NPSH. This will always be the case for any capacity smaller than that at which the "uncontrolled" head-capacity curve intersects the system-head curve, since in this region the level in the hotwell will be pumped down until the available NPSH will have been reduced to just equal the NPSH required by the pump, after which it cannot be reduced any further.

Thus, when a pump is operated under "submergence" control, the pump is operating in the break—i.e. cavitates—at all capacities less than "A." The pump may be more or less noisy at all capacities below "A," depending on a large number of factors involving design and operating conditions. However, the cavitation which occurs is generally not severely destructive in nature in the case of condensate pumps.

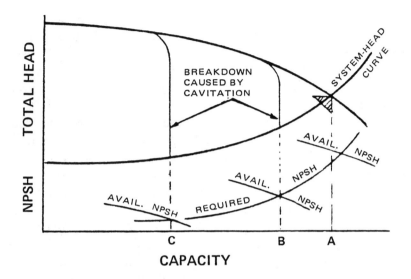

**Fig. 5-1. Condensate Pump Performance Curve
Showing How Capacity is Controlled by Means of
"Submergence" Control**

(Source: *Southern Power & Industry,* Feb. 1959)

To understand the effect of "submergence" control on pumps
operating with considerably higher values of NPSH, it is necessary to
analyze just what the mechanics of cavitation consist of. Cavitation
is a generic term which covers the entire scale of phenomena taking
place in and about an impeller when it is operated under conditions
where local pressures fall close to or below the vapor pressure of the
liquid pumped. The term covers not only the actual formation of
vapor bubbles, but also the limitation on capacity, the incidental
noise and the destructive effect on the impeller metal.

All these effects originate in the formation of vapor bubbles when
liquid pressure drops just below the vapor pressure. This is followed
by the recollapse of these bubbles to liquid when pressure increases
to a point just above the vapor pressure. The cause of cavitation
damage is the shock wave set up by the collapse or "implosion" of
these bubbles.

In a pump, these bubbles form at the inlet to the impeller when the available NPSH becomes equal to or slightly less than the required NPSH. The bubbles are carried along in the liquid stream as it passes through the impeller, until they reach a point where sufficient head has been generated to just exceed the vapor pressure. At this point, the bubbles collapse.

It is at this point where damage becomes evident. It cannot occur upstream of this point, since in that region there exists a mixture of liquid and vapor and the energy of the shock wave is dissipated in alternate compression and expansion of the vapor. But in the pure liquid phase, the shock wave is propagated through the liquid, until it is arrested by the surface of the impeller. If at the point of arrest, the shock wave possesses sufficient energy, it actually displaces a minute particle of material from the surface of the metal. Frequent and repeated occurrence of this process produces the pitting which is the usual symptom of cavitation. The degree of pitting for a given degree of cavitation is of course related to the relative "toughness" of the metal used in producing the impeller. In addition, there may take place a side-effect involving a chemical reaction of the metal to any gas held in solution in the liquid and liberated by the reduction in pressure.

That serious pitting occurs in pumps which cavitate at what may be referred to as "usual" levels of NPSH such as 15 or 20 ft is widely known. But it is a curious and much less known fact that condensate pumps operating on submergence control, work under essentially continuous cavitating conditions and yet do not usually suffer damage from this cavitation. There must, of course, be an explanation for this observed fact and the most logical explanation lies in the concept of "energy level" of the fluid stream. While this is by no means the only possible explanation for this phenomenon, and in itself may be only one of several contributing factors, it does lend itself to a rather ready explanation and is understandable without any detailed knowledge of pump design theory.

The energy level is indicated by the NPSH available. While this is generally expressed simply in feet of liquid, its actual units are foot-pounds per pound. In other words, the NPSH available is a measure of the available energy per pound of the liquid pumped. Thus, in a condensate pump operating at 3' NPSH, the collapse of the vapor bubbles occurs at an energy level of 3 ft-lbs per pound and the intensity of the shock wave of which we spoke is determined by this energy level.

In a pump operating at 15' NPSH, the energy of the fluid stream is five times as great and the intensity of the shock wave similarly greater. In view of this five to one ratio, it is not surprising that one of these instances causes no damage while the other one does. The difficulty, however, arises in establishing the limit of NPSH for non-destructive cavitation.

Despite the extensive literature on the subject of cavitation accumulated in the recent years, there is still a great scarcity of positive knowledge regarding the limiting factors of damage-free operation of centrifugal impellers under cavitating conditions. Most of the technical papers which treat this subject, deal with the fluid dynamics involved and develop various theoretical relations which—with the help of certain empirical coefficients—will yield approximate conditions under which cavitation may start. But whether operation under cavitating conditions will or will not have an abnormally destructive effect on the impeller, remains a matter of experience. We must consider that even the words "abnormally destructive" are subject to very wide interpretation.

What constitutes satisfactory life to one operator may be far from satisfactory to another. As a result, it becomes necessary to accept the fact that any limiting recommendations which may be presented here are strictly qualitative and subjective. They must be reevaluated by the reader in the light of his personal experience and opinion.

The limits are undoubtedly different in every installation because of minor differences in pump design, materials of construction, character of the liquid and general characteristics of the system in which the pump is operating. Admittedly, in a rather arbitrary fashion, it is common practice to select a limit of 5' NPSH for satisfactory life of a pump on submergence control. This limit may be considered as somewhat conservative, but will generally insure the installation against undesirable and unexpected difficulties.

In order to understand the problems involved in the control of condensate pumps, it is necessary to recognize that the capacity which any centrifugal pump will deliver into a system is determined solely by the intersection of the system-head curve and the head-capacity curve of the pump. Therefore, any variation in capacity must be obtained by varying either the pump head-capacity curve or the system-head curve. Since the delivered capacity of the condensate pumps must be capable of changing with the station load, means must be provided to either:

1) Change the system-head curve by throttling the discharge

2) Change the required pump capacity through the bypassing of excess over the desired net flow, or

3) Change the head-capacity curve by either varying the pump speed or by controlling the shape of this curve by suction limitations and operating the pump in a so-called "break."

When a closed feedwater system is employed, it is not permissible to limit capacity by operation in the break, since the condensate and boiler feed pump form, hydraulically, an integral unit and the capacity of each must be equal and is controlled by the feedwater regulator through the boiler feed pump.

If capacity control in an open system is exercised on the discharge side by means of a throttle valve, capacity variations are obtained by means of a family of artificially produced system-head curves (Fig. 5-2). Generally this throttle valve is actuated by a float or similar device in the condenser hotwell. The suction level is therefore maintained at a height somewhat in excess of the NPSH requirements of the condensate pump at all capacities.

In a controlled by-pass system, the pump operates at a relatively constant capacity and pressure. Net flow through the system external to the pump is varied according to the demand with the excess recirculated back to the hotwell. Since vent condensers require a constant flow of condensate regardless of the load station, this recirculation takes place downstream from them. Here again, NPSH requirements of the pump are always exceeded by the hot-well level.

However, this fact is not true in a system where no controlling discharge throttle valve or by-pass exists. Here the pump capacity is controlled by hot-well level variations since they vary the available NPSH. In effect, the pump is forced to follow load changes by means of its own hydraulic characteristics.

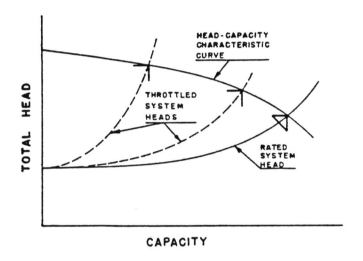

Fig. 5-2. Condensate Pump Performance Curve Showing
How Capacity is Controlled by Means of Artificially Produced
System-Head Curves through Throttling

Falling hot-well liquid levels which occur with lessening station loads and subsequent reductions in steam flow through the condenser, produce reduced available NPSH. Near saturation conditions usually exist, and NPSH requirements always increase with increased pump capacity. The interaction of these two conditions limit pump capacity to that value corresponding to existing hot-well level conditions (Fig. 5-3).

It should be noted that first stage impellers in pumps specifically designed for condensate and like services have much lower NPSH requirements than subsequent stages. They are also built to minimize the destructive effects of cavitation. This is not so in subsequent stages which are designed to produce other desirable hydraulic characteristics. Consequently, in extreme cases, damage through cavitation to the second stage impeller is frequently more severe than in the case of the first stage impeller.

The same sequence of events can occur in three, four, or five stage pumps to a varying degree depending upon which stages produce stuffing box pressures.

Naturally, the position and slope of the discharge system-head curve has a tremendous effect on the range of successful operation. In general, the higher the static head and the flatter the friction curve, the wider the range of possible operation without the danger of air leakage and attendant unsatisfactory operation.

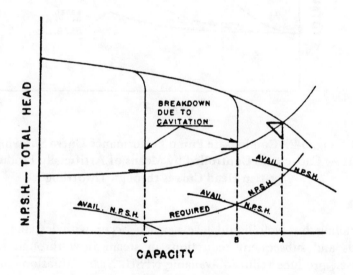

Fig. 5-3. Performance Curve of a Similar Pump Showing How Capacity is Controlled through Reductions in Available Net Positive Suction Head

DECREASING NPSH REQUIREMENTS*

The system designer has a number of ways in which he can attain the NPSH necessary for his pumps.

1) Use a double suction impeller.
   Whenever a double-suction impeller design is available for the desired conditions of service, it is likely to be the most desirable solution. Several benefits are based on the following considerations:

$$\text{Suction Specific Speed, S} = \frac{n_1 Q_1^{1/2}}{H_{sr_1}^{3/4}} = \frac{n_2 Q_2^{1/2}}{H_{sr_2}^{3/4}} \tag{a}$$

Where n = pump speed, rpm; Q = flow in gpm; H_{sr} = NPSHR where subscript$_1$ refers to a single-suction impeller and subscript$_2$ refers to a double-suction impeller

$$Q_2 = \frac{Q_1}{2}$$

i.e.: for double-suction impellers, Q_2 is half the total flow and for the same value of S, we can assume (a) $n_2 = n_1$ in which case $H_{sr_2} = 0.63 H_{sr_1}$ or (b) $N_{sr_2} = N_{sr_1}$ in which case $n_2 = 1.414\, n_1$.

Keeping the pump speed the same in both cases, as in Equation (a) we can reduce the NPSHR by 27 percent if we use a double-suction impeller (Fig. 5-4). Alternately, with a given NPSHR, as shown in our equation, we can operate a double-suction pump at 41.4 percent higher speed (Fig. 5-5). In both cases the value of S, the suction specific speed, remains the same.

2) Use a larger impeller eye area.
   In this case, the idea is to decrease the NPSHR by reducing the entrance velocities into the impeller. Within limits, the practice is feasible; these lower velocities will probably not

*Pump World, Worthington Co., 1980, Vol. 6, No. 1.

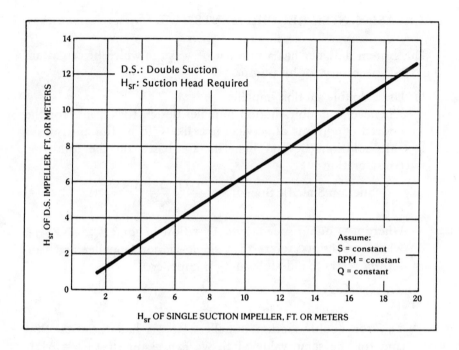

Fig. 5-4. Effect of Using a Double-Suction Impeller on Required NPSH

(Source: *Pump World,* Worthington Pump Co.)

affect pump performance at or near the best efficiency point. When such pumps run at part capacity, however, reduced velocities can lead to noisy operation, hydraulic surges, and premature wear: symptoms of internal recirculation at the impeller suction.

3) Raise the liquid level.
This is the simplest solution—when it can be done. Of course, there are times when this is not practical. The liquid level may be fixed, as in the case of a river, pond, or lake; the amount by which the level would have to be raised may be completely

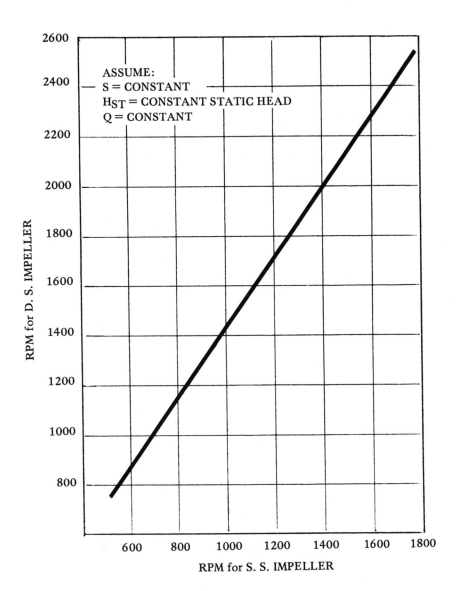

ASSUME:
S = CONSTANT
H_ST = CONSTANT STATIC HEAD
Q = CONSTANT

**Fig. 5-5. Effect of Using a Double-Suction Impeller
on Permissible Speed**

(Source: *Pump World,* Worthington Pump Co.)

impractical; or the cost of raising a tank, a fractioning tower, or a deaerating heater may be excessive.

Nevertheless, this approach should be considered. Sometimes just a few extra feet of suction head may permit the selection of a much less expensive or much more efficient pump. The resultant savings in first cost, energy, or maintenance may far outweigh the additional costs incurred by raising the source of the liquid.

4) Lower the pump.
Just as in the case of raising the liquid level, the cost of lowering the pump may be a wise investment, since it may permit the selection of a higher speed, less costly, and more efficient pump. If it's not practical to lower the floor level at which a horizontal pump is installed, the pump user can try a vertical pump operating in a sump. If an open sump is impractical—as in the case with condensate pumps or pumps handling volatile, inflammable, or toxic liquids—the solution may be a vertical submerged (barrel) pump, mounted in its own closed sump: a tank or "can" which is sunk into the floor. The pump is simply made long enough to reach the level where sufficient NPSH is available for the first-stage impeller. The diameter and length of the can must allow proper flow between the pump and the tank, and for the turn and flow into the bellmouth of the pump.

5) Use slower speeds.
In weighing the relative adequacy of a particular pump design from the point of view of required NPSH values, it is most practical to make comparisons of the suction specific speed that it represents. S, suction specific speed, has been defined previously.

The NPSH curves recommended by the Hydraulic Institute are based on S values from 7480 to 10,690 with most of the curves falling below 8500. These values are relatively conservative and may be raised somewhat, but values of 8500 to 9000 should NOT be exceeded, particularly if the pump is required to operate over a fairly broad range of capacities.

It is obvious from our suction specific speed equation that a lower speed pump means lower required NPSH. The problem, however, is that a lower speed pump is more expensive than a higher speed pump designed for the same conditions of service. More frequently, it is also less efficient.

6) Use an inducer.

To achieve low NPSH requirements, the pump is fitted with an inducer and a high-solidity impeller. The high-solidity impeller has three to five times the normal number of radial blades. However, the interesting part of this design is the inducer. Many pump companies now offer an inducer to reduce NPSH requirements significantly. Conventional single-stage, high-horsepower centrifugal pumps can require substantial NPSH because of their high rotating speeds. Inducers, which achieve considerably higher suction specific speed than impellers, are used to reduce the NPSHR. This will be discussed at greater length in a subsequent section.

7) Use an oversized pump.

About the only harm that oversizing does for the "average" small, nominal-head pump is to move the operating range to the flat part of the H-Q curve. This bothers some people. Yet, even if the curve is slightly unstable at low flows, these average pumps usually give no trouble.

Not the same situation for oversized high-head pumps. To the dangers of pulsations or surges due to rotating-stall or unstable H-Q curve, add the probability of cavitation noise and cavitation erosion, at *normal* flow as well as low flow. The oversizing may cause the rated capacity to be below 60 percent of the b.e.p. capacity for that size pump! And if the oversized high-energy pump has a double-suction impeller, add the possibilities of axial shaft shuttling below 50 percent Q, making for early failures of bearings or mechanical seals!

What should pump application engineers do to avoid partial-flow cavitation on high-head pumps? If it becomes necessary to use an oversized pump for NPSH reasons, give more thought to the provisions to assure a higher minimum flow than usual. Ask the manufacturer to distinguish between minimum for temperature-rise protection, and minimum for zero cavitation damage, or whatever.

8) Subcool the liquid.
Available NPSH may be defined as

$$H_{sa} = Z + (P_s - P_{vp}) - (h_{fs} + h_i)$$

Where H_{sa} = NPSH$_a$
Z = static head
P_s = pressure above liquid level
P_{vp} = vapor pressure of the liquid
h_{fs} = friction loss in suction piping
h_i = entrance loss at pump inlet

and by definition is the energy in excess of the vapor pressure of the liquid at the pumping temperature.

All the methods examined up to now have *increased* NPSH by increasing the static component Z or *reduced* NPSHR in various ways. There is one other possibility: NPSHA may be increased by *decreasing* the liquid's vapor pressure, which is done by reducing the pumping temperature.

The most obvious way is to inject cold liquid ahead of the pump suction as illustrated in Fig. 5-6.

A typical installation using this system might be a steam-electric power plant. The colder injection flow is taken from the upstream side of the feedwater cycle, ahead of the deaerating heater. Of course, introducing nondeaerated condensate in the boiler feed pump suction is not recommended, especially in the case of high-pressure boilers where the possible effects of oxygen contamination may be severe.

A more highly recommended practice is to eliminate all possibility of oxygen contamination by installing a heat exchanger in the suction piping to subcool the deaerated feedwater. Friction loss through the heat exchanger would have to be added to the desired increase in NPSHA. To avoid wasting heat in subcooling, the cooling medium can be the condensate itself on the way to the deaerating heater.

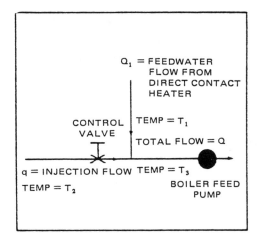

Fig. 5-6. Subcooling by Injection of Colder Water

9) Use a booster pump.

The principle of the booster pump is simple. A low-speed, low-head pump, generally of single-stage design, is installed ahead of the main pump to provide it with greater NPSHA than can be made available strictly from static elevation of differences.

Booster pumps are often used ahead of high-pressure multistage boiler feed pumps. Modern practice uses operating speeds well above 3600 rpm for the main feed pumps, so NPSHR values may run as high as 150 to 250 feet. It is seldom practical to install the deaerating heaters from which the boiler feed pumps take their suction at an elevation high enough to meet such requirements. On the other hand, single-stage, double-suction booster pumps, operating at the lower speed of 1750 rpm may require as little as 25 to 35 feet of NPSH, neatly solving the problem.

Boosters not only "get hold of" the liquid at much lower NPSH than the main pumps, but also make far less noise and trouble if they themselves cavitate at times. Remember the lower pressure energy in the booster's impeller to collapse cavities.

Boosters should be selected to provide *at least twice* or three times the NPSHR required by the main pump! Remember, the NPSHR of the main pump is a 3 percent head-drop performance impairment—by definition. The main pump, if handling water liquids, will suffer from cavitation *erosion* at partial flows. It could need more than three times NPSHR3PC* to prevent cavitation damage, when operating "off design"— below 60 percent b.e.p.

If a booster's total head gets up near 200m (650 ft), however, be cautious. The booster itself may become a high-energy pump, vulnerable to cavitation erosion when "off design."

10) Use variable speed operation.
Partial flow operation at reduced speed is usually a satisfactory way to protect a high-energy pump from rapid cavitation erosion. When starting up nuclear power plants, the feed pumps may run many months at extremely low flow. Those that have variable-speed drives have a great advantage over constant-speed pumps.

However, variable speed capability does not ensure that pumps will be kept from operating at low percentages of b.e.p. flow. For some applications, operators frequently run two pumps instead of one, both at maximum speed, to get the most flow into a system.

---

*NPSHR3PC refers to the condition where the pump delivery is reduced by 3 percent.

Unless the system can take nearly twice the flow of one pump, both pumps could suffer excessive cavitation erosion. Each would be operating at 40 to 70 percent of the b.e.p. flow at the higher speed. It would be better for the impellers to run both pumps at slightly reduced speed. Next best would be to run one pump at maximum speed.

11) Use two-stage pumps.
At one time there was quite an efficiency advantage for two-stage, to offset the higher cost and greater number of parts compared to single-stage. Then single-stage designs were extended to larger pumps, high heads from 600 to 2000 ft (600m), higher speeds, with efficiencies 80 to 89 percent. Now pump sellers have become reluctant to offer two-stage pumps whenever single-stage designs are available. The pendulum may have to swing back soon. Pump manufacturers are not usually responsible for corrosion and erosion in field operation. But, cavitation erosion may have to be considered separately from erosion by solids; it is becoming such a problem, and two-stage pumps have much lower head-per-impeller, and consequently need reduced NPSH.

PUMPING BOILING FLUIDS

Boiler feedwater pumps are a particular example of a pump dealing with fluid which can readily flash. The pumping of a liquid close to the boiling point pressure has always been a problem. The incipient formation of vapor, or a loss of head, is the advance indication of trouble. This is the pump "vapor barrier," so to speak—the "never never land" between a liquid and its vapor. Centrifugal pumps merely approach it. When they do, they suffer possible cavitation damage or loss of operation entirely.

The generally accepted centrifugal pump parameters, such as net positive suction head (NPSH) and suction specific speed (S), are being used increasingly as standard references in studies of cavitation effects, relative pump-suction performance, design, and application of criteria.

It is becoming recognized, however, that the application of these parameters should be approached with caution. If, for instance, a pump parameter rating is to be established from tests on liquids of unusual thermodynamic characteristics, such as cryogenic fluids or petroleum, and the like, it is necessary to apply correction factors to make the rating compatible with the usual understanding of the term.[1]

It is also evident that these parameters may cause confusion in the understanding.

If, for instance, a liquid is boiling, the acceptable application standard of NPSH[2] implies that, regardless of the pressure above the surface of the boiling liquid, this surface level must be located above the pump inlet in order to be pumped successfully. Standard practice, therefore, dictates that a positive static head must be provided on all boiling-liquid applications so that the liquid will flow by gravity into the pump. The possibility that a boiling liquid can be "lifted" by the pump to its inlet is not even inferred. The force of precedent would at first classify this as impossible since it would represent an available net positive suction head that was a *negative* value! This is *below* the zero limit of the NPSH application parameter and *infinity* if applied to the suction specific speed S-parameter.

These established concepts make it difficult, without considerable explanation, to accept and understand initially the performance of a pump that can pump under these unorthodox conditions.

In centrifugal pump practice, it is considered necessary to provide a higher inlet absolute pressure on the liquid than its saturated-vapor pressure. This is to compensate for losses inside the pump and thus avoid the formation of vapor therein. Should vapor be present, pumping is difficult, if not impossible.

A centrifugal pump is also vulnerable to the presence of air or gas already in the liquid or entering at its suction. This air or gas, however, is a finite entity which can be dealt with by various familiar devices to "prime" the pump or bodily remove the air or gas.

Vapor, however, is elusive. It flashes into existence as a result of temperature rise or pressure reduction and it disappears just as

[1] A. J. Stepanoff, ASME Paper No. 63-AHGT-22.
[2] Standards of the Hydraulic Institute B-49.

quickly by condensing. Because of this quality, the conventional practice when pumping liquids close to the boiling point is to prescribe means that will prevent vapor reaching the pump. Such devices include traps, flash tanks, elevated receivers with start-stop controls, jets and cooling equipment, or booster pumps ahead of the higher pressure pumps.

A "multiphase" pump has been proposed as an answer to the problem. It combines a liquid ring first stage with a more conventional succeeding stage. The operation of the liquid ring portion of the pump is shown by Fig. 5-7.

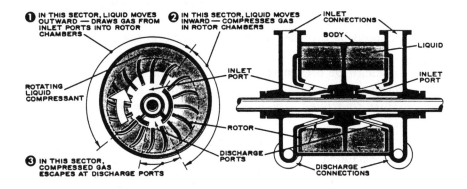

Fig. 5-7. Operation of Typical Liquid Ring Pump
(Source: Nash Pump Co.)

A liquid-ring pump and a centrifugal pump may thus be closely combined in an unusual staged design. This unit will continuously pump fluids in their liquid, vapor or gas phases or in any combination of these, as typical in two-phase flow.

The functionally selective staged design of the Nash pump simultaneously extended the unit's overall fluid handling characteristics to include performance in areas impossible for each element when operating by itself.

This new pump has demonstrated characteristics that are novel and useful on applications involving condensate and air removal from vacuum driers, steam-heating devices, evaporators and condensers. Its value will become apparent in many other vapor/liquid recovery services but its initial acceptance will be somewhat handicapped by the fact that these novel characteristics are beyond the limits of established pumping practice and parameters.

ENTRAINED AIR

Air or gas enters a pump in a variety of ways: leaking at joints in suction piping, at valve packings, and especially from vortices which readily form in the vicinity of pump suctions, especially in shallow pump basins. Stuffing boxes on pumps having a below atmospheric suction should be provided with water injection for cooling, sealing and lubrication.

If gas cannot be prevented from entering the pump suction, judicious venting may help. This can be in the form of an automatic device, float controlled, which will prevent the pump from becoming vapor-bound.

In general, air or gas is undesirable in a centrifugal pump and should be excluded if at all possible. If a limited amount of gas is handled, then caution should be used in selecting a pump. Most standard centrifugal pumps handle 3 percent gas without difficulty. Many handle more than this, but with varying degrees of uncertainty. Where more than 7 percent gas is handled, a self-priming or other special pump should be used unless the excess gas is vented off before it reaches the pump.

Pumps may be modified to handle greater amounts of air by impeller design, or by angle cutting the tips of the impeller vanes. However, these modifications will result in loss of capacity and head. Self-priming pumps, jet pumps, or liquid ring pumps in combination with centrifugal stages are possible.

Sealing liquid for packings which operate at vacuum conditions should be clean and injected at about 15 pounds pressure. Clean liquid is available by utilizing some of the discharge pressurized liquid, cleaning with a small centrifugal separator, and then injecting the clean water into the seal. A filter may be provided to similarly clean up the pump discharge, for injection into the seal. Of course, if potable water is available, it is ideal for seal flushing.

PRIMING CENTRIFUGAL LIFT PUMPS

Centrifugal pumps which are required to lift water work satisfactorily once flow has been established. However, such pumps, when starting up after a period of shutdown, will frequently be dry. Foot valves are commonly provided to hold prime in a pump, but they may be unsatisfactory for two main reasons: they leak, losing prime, or they impose a restriction on the suction. One method of insuring that the pump prime does not leak out is to provide a small bleed line from another pump or from potable water, to keep the suction pipe full.

A more satisfactory method is to connect the pump casing to a vacuum system, in which a continuous vacuum is maintained sufficient to hold water in the pump casing. For this vacuum, a liquid ring pump is most satisfactory. If energy savings are desired, an arrangement can be provided wherein the vacuum pump starts when the main pump is to be started. Once the prime is established, the main pump is allowed to start and the vacuum pump is shut down.

VORTICES

Vortices at pump intakes are potential sources of a variety of pump operational problems. These are: (1) reduced pump efficiency; (2) vibration and noise; (3) undue wear of bearings; and (4) accelerated deterioration of impeller blades due to abrasion by entrained debris, corrosion due to excessive air, and pitting. Prevention of vortices is thus a major criterion in the hydraulic design of a sump. Since analytical solutions to the problem are almost impossible, greater reliance and emphasis have come to be placed on empirical methods involving physical models.

Efforts have been made in the past to develop a few simple design rules based on experience. Examples of such design rules may be found in the Hydraulic Institute Standards and the more recent publications of the British Hydromechanics Research Association (BHRA). Application of these rules yields a conservative, standard sump design. However, departures from standard designs are often inevitable due to constraints of site and structure, or are desirable for economic reasons, and without the aid of a model study, it is difficult to assure the soundness of a proposed modified design.

Pull the plug in a sink full of water, and a vortex forms. This rotation is present in all real flow situations due to the velocity profile of the fluid moving toward the outlet. Any changes in flow direction, abrupt changes in boundary geometry, or flow patterns generated by multiple pumps will generate eddies which augment incipient vortices. A vortex may be first noticed as a dimple in the water surface, which gradually becomes deeper until its axis approaches the pump intake. All the vortex filaments converge, just as stream lines do, and by virtue of the property of conservation of rotational energy, the vortex forms. It becomes obvious that the closer the water surface is to the suction bell, the easier it is for the vortex to form.

Conditions that are favorable to vortex development are large velocity in the intake pipe, low depth of submergence of the intake or suction bell, and presence of low momentum fluid or dead zones. These factors together with approach channel geometry form the most important elements to be considered in the design of a sump.

To limit the severity of vortexing to within permissible limits, the following basic principles have been established:

1) Ensure proper approach flow. Wherever possible a straight approach channel is to be preferred. Flow should be uniform across the width of the channel.

2) Provide submergence of at least 1.5D, where D is the diameter of the suction bell, for both horizontal and vertical suction bells. Both NPSH and vortex prevention should be considered in establishing the actual submergence. The larger of the two should be used.

3) Keep velocity in the approach channel at about 0.5 to 1.0 fps.

Very small approach velocities not only make the sump uneconomical, but also are not effective in vortex prevention. The low momentum of the approach flow is often found to be conducive to the development of a vortex rather than to its prevention. (Where the flow has appreciable sediment content, very low approach velocity may cause it to settle in the sump).

4) Avoid dead or stagnant zones of liquid.

5) In the case of multiple pumps arranged in a row, align them normal to the approach flow.

6) Provide individual sumps for each pump, if feasible.

7) Streamline obstructions such as pillars and support piers so that they do not shed large vortices, and do not locate them too close to the inlet.

8) Screens placed in the approach flow also help to make the flow more uniform, but partially clogged screens could distort the flow. In one application, deep trash racks were found to be very effective as flow straighteners.

Where it is not possible or feasible to provide proper approach conditions, such as in an existing sump, it may be necessary to install antivortex devices. Some devices which have been successfully used are floating balls on the surface which reduce surface energy, a baffle plate installed as shown by Fig. 5-8, a vertical splitter plate mounted on the sump floor, suction bells with turning vanes, a floating raft on the surface above the suction.

The raft should be of ample dimension to counteract the possibility that vortices may become started at some distance away from the axis of the hole at the bottom of the tank. It is also advisable to provide means to prevent the rotation of the float by anchoring it in some manner, while still permitting it to float freely on the surface of the water.

Frequently, screens or trash racks ahead of the pump suction will minimize the vortex action. A stabilizer/separator, such as shown in Fig. 5-9, also has a beneficial effect in improving suction conditions.

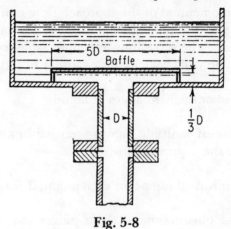

Fig. 5-8

(Source :*Power & Fluids,* Worthington Pump Co.)

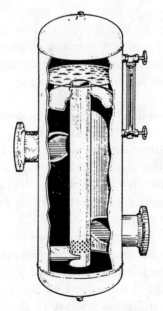

Fig. 5-9. Stabilizer Separation

(Source: Fluid Kinetics Co.)

6

A Yardstick for NPSH Requirements

*NPSH is net positive suction head. In its simplest form, it is pictured in Fig. 6-1. Pure water is being pumped at 212°F. The atmosphere is 14.696 psia, which is the boiling point of pure water at 212°F.

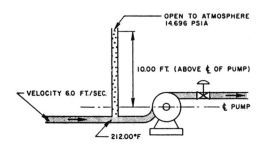

OPEN TO ATMOSPHERE
14.696 PSIA

10.00 FT. (ABOVE ₵ OF PUMP)

VELOCITY 6.0 FT./SEC.

₵ PUMP

212.00°F

Fig. 6-1. NPSH in Its Simplest Form

The flow has been gradually increased until there is some cavitation or the head curve of the pump shows a detectable drop. (Thus defining NPSHR.)

A glass tube rising vertically from the suction line, open to the atmosphere, shows the water level standing 10 feet above the center

*by A. Whistler. Abstracted from *Petroleum Refiner,* January 1960.

line of the pump. At some spot in the pump, the pressure has dropped to 14.696 psia and the water is vaporizing slightly.

All of the 10 feet of static has been used up at some point inside the pump. But, in addition, the pump has used up a little more than this. The liquid in the suction line has energy of motion, called kinetic energy. This has been used also. The formula for this energy, in feet, is

$$\text{Kinetic Energy} = \frac{(\text{Velocity})^2}{2 \times g} = \frac{V^2}{64.4}$$

Since the water is moving at 6 ft/sec, its kinetic energy is

$$\text{KE} = \frac{(6)^2}{64.4} = 0.56 \text{ feet}$$

The NPSH required is, therefore:

$$10 + 0.56 = 10.56 \text{ feet}$$

The case just described is the simplest. To cover all cases, the definition has to be a little more technical.

GENERAL DEFINITION

NPSH is the total net energy (static and kinetic) in the suction pipe, above the vapor pressure of the liquid, required to maintain pump total head. Reference point is usually center line of the pump shaft (for vertical pumps it is usually center line of suction). Sometimes a certain deviation is allowed from the total head.

The NPSH is computed as follows:

1) Read pressure in suction line—convert to feet of liquid.

2) Compute velocity in suction line.

$$\text{Kinetic Energy (ft)} = \frac{V^2}{64.4}$$

3) Add 1) and 2) to obtain gross suction energy.

4) From temperature in suction line, obtain vapor pressure and convert to feet of liquid.

5) Subtract 4) from sum of 1) and 2).

6) Correct for difference in elevation of center line of suction and center line of pump. If suction line is lower, you subtract.

All this may appear complicated, but it is used in the customary factory test. Fig. 6-2 shows the set-up. The pump to be tested takes suction from a tank or pit. The pump and piping are filled with water—the foot valve prevents runback to the pit.

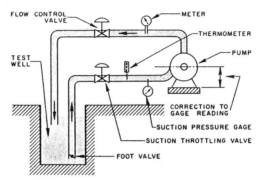

Fig. 6-2. Standard Factory Test Set-Up to Find NPSH

A series of points of various flow rates are run. Say, 300 gpm is one such point. This flow is held constant by the flow control. The suction valve is gradually closed until there is some evidence of cavitation. It might be by sound or it might be by a drop in the capacity curve (the capacity curve has already been run with wide open suction).

The following readings and computations are made:

1) Suction temperature 80°F, conversion factor
   psi to feet of water   = 2.314 ft/psi
   Vapor pressure from steam tables
   0.5068 psia × 2.314   = 1.173 ft
2) Suction gage 5.58 psia × 2.314   = + 12.92 ft
3) Suction gage is 6 inches below
   center line of pump   − 0.50 ft
4) Pressure in suction at pump center
   2) + 3)   = 12.42 ft

5) Velocity of liquid, 4-inch line,
    300 gpm   =   8.025 ft/sec

Kinetic energy = $\dfrac{(8.025)^2}{64.4}$ = + 1.00 ft

6) Gross energy in suction 4) + 5)   13.42 ft

7) Less vapor pressure   − 1.17 ft

8) NPSH 6) − 7)   12.25 ft

From a theoretical point of view, the methods of Figs. 6-1 and 6-2 can be considered equal. But there are some very important practical differences. In Fig. 6-2, the water contains dissolved air, but in what degree of saturation no one knows. However, we do have considerable data on pumps tested by Fig. 6-2.

Data by Fig. 6-1, with pure deaerated water, appear to be completely missing. There are a few data taken on hot water, but the tank was still exposed to air. Undoubtedly much of the air was eliminated, but it takes more than simple heating to remove it. Look at the design of deaerators for boiler feed water which still leaves a trace.

Tests are run occasionally where vacuum is applied to the suction tank. Undoubtedly, this also removes a large percentage of the air, but certainly not completely.

Unfortunately it is human to think of the perfect method when we actually have data taken from an imperfect, though very practical, method. It might help to use more exact names. Call data from Fig. 6-1 *Absolute NPSH,* and from Fig. 6-2 *Open Cold Water NPSH.*

NPSH REQUIREMENTS FOR OTHER FLUIDS

For years there has been a widespread belief in the petroleum industry that "hydrocarbons require less NPSH than water." Several types or methods of determining NPSH corrections for hydrocarbons have been used. Several pump manufacturers have used a straight multiplier for all hydrocarbons.

Fig. 6-3 shows a performance curve on this basis—quoted sometime ago for an oil pump. One curve of NPSH is for water, presumably *Open Cold Water*. A second is marked *Oil*. This is an almost
constant fraction (0.6) times the water curve. No qualities or conditions for the oil, such as pressure, temperature, gravity, etc., are
specified.

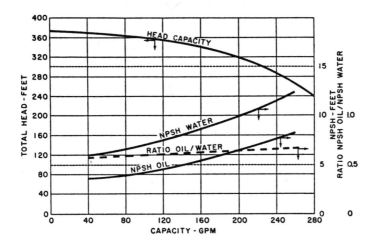

**Fig. 6-3. This Performance Curve Was Quoted Sometime Ago
for An Oil Pump**

Actually, Fig. 6-3 is for a pump taking hot lube-oil from a vacuum
tower. But pumps for other duties, such as cold gasoline at atmospheric pressure, have about the same ratio between NPSH for oil
and for water.

However—perhaps luckily—it is doubtful that such oil NPSH curves
have been used very often in actual installations. Certainly the pump
manufacturers have never claimed that these oil NPSH curves were
anything more than what they are, but such curves are potentially
dangerous. Some engineer who doesn't know the background of
these curves might use them. They should be eliminated from standard performance charts.

Fig. 6-4A shows an early Hydraulic Institute chart used by Whistler in this discussion. However, a better method is now suggested by the standards of the Hydraulic Institute. Fig. 6-4B is a reproduction of the chart.

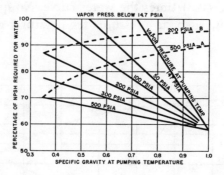

Fig. 6-4A. Early Hydraulic Institute Curves for NPSH Correction

The following limitations and precautions should be observed in the use of Fig. 6-4B.

Until specific experience has been gained with operation of pumps under conditions where this chart applies, NPSH reductions should be limited to 50 percent of the NPSH required by the pump for cold water.

This chart is based on pumps handling pure liquids. Where *entrained* air or other noncondensable gases are present in a liquid, pump performance may be adversely affected even with normal NPSH available (see below) and would suffer further with reductions in NPSH. Where *dissolved* air or other noncondensables are present, and where the absolute pressure at the pump inlet would be low enough to release such noncondensables from solution, the NPSH required may have to be increased above that required for cold water to avoid deterioration of pump performance due to such release.

For hydrocarbon mixtures, vapor pressure versus temperature relationships may vary significantly with temperature, and specific vapor pressure determinations should be made for actual pumping temperatures.

In the use of the chart for high temperature liquids, and particularly with water, due consideration must be given to the susceptibility of the suction system to transient changes in temperature and absolute pressure, which might necessitate provision of a margin of safety of NPSH far exceeding the reduction otherwise available for steady-state operation.

Because of the absence of available data demonstrating NPSH reductions greater than ten feet, the chart has been limited to that extent and extrapolation beyond that limit is not recommended.

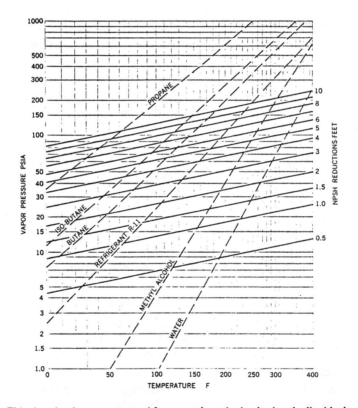

NOTE: This chart has been constructed from test data obtained using the liquids shown.

Fig. 6-4B. NPSH Reductions for Pumps Handling Hydrocarbon
Liquids and High Temperature Water

(Source: Hydraulic Institute Standards)

Instruction for Using Chart for Net Positive
Suction Head Reductions (Fig. 6-4B)

Enter Fig. 6-4B at the bottom of the chart with pumping temperature in degrees F and proceed vertically upward to the vapor pressure in psia. From this point follow along or parallel to the sloping
lines to the right side of the chart, where the NPSH reductions in feet
of liquid may be read on the scale provided. If this value is greater
than one-half of the NPSH required on cold water, deduct one-half
of the cold water NPSH to obtain corrected NPSH required. If the
value read on the chart is less than one-half of the cold water NPSH,
deduct this chart value from the cold water NPSH to obtain corrected
NPSH required.

Example: A pump that has been selected for a given capacity and
head requires a minimum of 16 feet NPSH to pump that capacity
when handling cold water. In this case the pump is to handle propane
at 55°F, which has a vapor pressure of 100 psia. Following the procedure indicated above, the chart yields an NPSH reduction of 9.5
feet, which is greater than one-half of the cold water NPSH. The corrected value of NPSH required is therefore one-half the cold water
NPSH or 8 feet.

Example: The pump of example above has also been selected for
another application to handle propane at 14°F, where it has a vapor
pressure of 50 psia. In this case, the chart shows an NPSH reduction
of 6 feet, which is less than one-half the cold water NPSH. The corrected value of NPSH is therefore 16 feet less 6 feet, or 10 feet.

Some variables are not included directly in Fig. 6-4A, such as 1)
vapor density, 2) latent heat, 3) specific heat. Some of these go along
to some extent with pressure. At any rate, reasonable results would
be obtained for the low gravity region. However, the influence of
specific gravity is in reverse to what it should be. A better form of line
is demonstrated by lines A and B. These are plotted on the assumption
that the 0.35 specific gravity points would be correct. The results
based on Fig. 6-4A, may be compared to those based on Fig. 6-4B, to
indicate how information is developed.

Actually, the dangerous area is at high specific gravity. For example, a heavy crude being pumped from a 5 psig (20 psia) receiver

after the crude has been flashed from a higher pressure would read 85 percent of water NPSH. This would invite trouble.

One pump manufacturer for a long time used a chart that is a variation of Fig. 6-4A. The main difference is that at 14.7 psia, and below, the reading was 100 percent.

MATHEMATICAL ANALYSIS

A strictly mathematical analysis of the problem will not be made. If the mathematics are difficult, concentrate on understanding 1) the fundamental picture, 2) the assumptions made, and 3) the final chart. For mathematical derivation, see page 97.

The fundamental picture is shown in Fig. 6-5. The picture at the top shows liquid flowing in a converging pipe from left to right. Pressure at Point A is well above the bubble point. At Point B, the pipe starts to neck down, hence, the velocity starts to increase. Pressure has to drop in order to supply the added energy of velocity (increase of kinetic energy). This continues until, at Point C, the pressure has dropped to the bubble point. Between C and D, more and more vapor is formed.

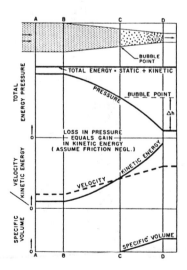

Fig. 6-5. Liquid is Flowing in a Converging Pipe

The temperature remains constant from A to C. Then, the vapor formation requires heat (latent heat). Assume that this heat comes from the liquid. The only way the liquid can give up heat is by a drop in temperature (sensible heat). Of course, this is not *exactly* true. But remember only a very small *weight* percentage of the liquid turns to vapor. Hence, heat demand is small. It is all well within the accuracy needed.

The eye of the impeller is somewhere near Point D. The region beyond Point D is far enough inside the impeller so that the pump is supplying all the added energy. None comes from the pressure in the suction line. The vapor will liquefy as the pressure goes up. At any rate, Point D is the point of maximum vapor.

The drop in pressure between Points C and D is called Suppression Head and is represented by the symbol Δh. This is customarily in feet of liquid because it puts all liquids on a common basis (equal Δh give equal increase in velocity for all liquids).

$$\Delta h = \left(\frac{144}{\rho_1}\right) \left(\frac{P_1 - P_2}{T_1 - T_1}\right) (V) \left(\frac{\rho_v}{\rho_1}\right) \left(F + \frac{H}{100 \, C_p}\right) \qquad (1)$$

Δh = Suppression in pressure below initial bubble point in ft of liquid.

F = Slope of initial vaporization, °F/wt. percent vapor. For example, a mixture might start to vaporize at 100°F and be completely vaporized at 200°F. Then F = (200−100/100) = 1.0. For pure components F = 0.

H = Latent heat, Btu/pound.

C_p = Specific heat of liquid.

V = Actual cubic feet of vapor formed per 100 cubic feet of liquid.

$\dfrac{P_1 - P_2}{T_1 - T_2}$ = Slope of vapor pressure curve, psi/°F, taken at point of operation.

ρ_v = Density of vapor, pound/cubic feet.

ρ_1 = Density of liquid, pound/cubic feet.

If you examine equation (1) carefully, you will see that for any set point of pressure and temperature, it boils down to:

$$\Delta h = a \text{ multiplier times V} \qquad (1a)$$

This means that, if we double V, we get double the value for Δh. The value of V must always be known to solve for Δh.

How much vapor can a pump stand? There isn't much to go on; however, there is a little. Some engineer with a pump company told me that they had injected air in a pump suction up to 3 cubic feet per 100 cubic feet of water without noticeably harming performance. Other pump engineers have agreed that this is a reasonable figure.

It is agreed that such information is not scientific. But suppose we pick 2 cubic feet per 100 cubic feet of liquid. All the pump engineers I've talked to think that this is quite safe. To be sure, different pumps must act differently, and the limit must change throughout capacity for any one pump. But it will be some time before we get such data.

To show how all this might be applied, look at an example. Suppose we plan to pump pure propane at 200 psia. From Fig. 6-6A, we read Δh= 3.2 feet.

The drum from which we will pump has a liquid level 10 feet above the centerline of the pump (after deductions for line friction).

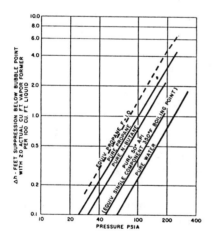

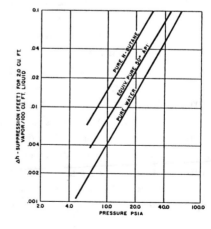

Fig. 6-6A *(above).* **Computed Suppression Below Bubble Point for 2.0 cu ft Vapor per 100 cu ft Liquid. High Pressure Range**

Fig. 6-6B *(right).* **Low Pressure Range**

For zero vapors at the eye, the NPSH needed would be 10+3.2= 13.2 feet. The factory will test on cold water. On Fig. 6-6B, the lowest reading for water at 4.5 psig is Δh=.001 feet for 2 cubic feet of vapor at the eye. This corresponds to about 150° F. The test will certainly be run at 100° F or less.

The correction for water is infinitesimal. Therefore, we would specify that the pump must not require more than 13.2 feet of NPSH under water test.

We should also call attention to the fact that we have said nothing about damage to the impeller from cavitation. That is another problem. Actually, most engineers would not plan on allowing any vaporization at design load. But most of them put a fair "pad" in the specified capacity. Therefore, the pump manufacturer quotes an NPSH that already has a factor of safety. Hence, the formula might be used to prevent doubling the factor of safety.

Another use is in analyzing data or in trouble shooting. I have used this formula several times to throw out some erroneous conclusions.

Fig. 6-6 gives the computed suppression when 2.0 cubic feet of vapor are formed per 100 cubic feet of liquid. If other values of vapor are desired, simply multiply by the ratio of V/2.0.

Fig. 6-6 is probably most useful in clearing up a few ideas we have had. In the first place, the correction is a straight subtraction and *not* a multiplier, as supposed. Nor do the various hydrocarbons have the same value at the same pressures. We expect water to have a different value, which it does.

Notice also the extremely small suppressions at low pressures. Even if we assume enormous amounts of vapor, the computed suppression is not measurable. Yet open cold water tests often show a small amount of drop in the total head curve before the final sudden drop. This is a proof of dissolved air, I believe.

Some have advanced the idea of what might be called delayed vaporization, a sort of supersaturation. Yet cavitation tests on models, using water, show that the bubbles form and collapse in a few thousandths of a second.

Of course, the same tests are not reported on hydrocarbons, but we know that where hydrocarbons at bubble point pass through a throttling valve, enough vapor forms in one one-hundredth second to form frost on the valve and also to affect the valve capacity.

DERIVATION OF FORMULA

Start with the case for a pure liquid (one component). Fig. 6-7 shows an enlarged section of the vapor pressure curve. Assume a straight line between points A and C. This isn't as inaccurate as might appear. Usually the temperature drop between A and C is not more than $5°F$—sometimes only $1°F$.

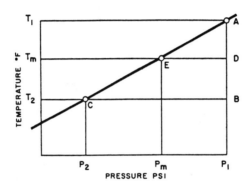

Fig. 6-7. Vapor Pressure Curve for Pure, One Component Liquid

Fig. 6-7 ties in with Fig. 6-5. Point A of Fig. 6-7 corresponds to Point C of Fig. 6-5. Point E of Fig. 6-7 corresponds to Point D of Fig. 6-5.

In Fig. 6-7, triangle ADE is similar to triangle ABC which means that

$$\frac{P_1 - P_2}{T_1 - T_2} = \frac{P_1 - P_m}{T_1 - T_m} \tag{2}$$

The heat used up in vaporization is

$$V\rho vH \tag{3}$$

And this is assumed as coming from sensible heat of the liquid

$$100\rho_1 (T_1 - T_m)C_p \tag{4}$$

Therefore

$$V\rho vH = 100\rho_1 (T_1 - T_m)C_p \tag{5}$$

Writing (5) in different form

$$T_1 - T_m = \frac{V\rho vH}{100\,\rho_1\,C_p} \qquad (6)$$

Substituting (6) into (2)

$$\frac{P_1 - P_2}{T_1 - T_2} = \left(P_1 - P_m\right)\left(\frac{100\rho_1\,C_p}{v\rho vH}\right) \qquad (7)$$

$$\text{But } \Delta h = \left(P_1 - P_m\right)\left(\frac{144}{\rho_1}\right) \qquad (8)$$

$$\text{or } \left(P_1 - P_m\right) = \frac{\Delta h \rho_1}{144} \qquad (8a)$$

Substituting (8a) into (7)

$$\frac{P_1 - P_2}{T_1 - T_2} = \frac{\Delta h \rho_1}{144}\left(\frac{100\rho_1\,C_p}{v\rho vH}\right) \qquad (9)$$

$$\text{or } \Delta h = \left(\frac{144}{\rho_1}\right)\left(\frac{P_1 - P_2}{T_1 - T_2}\right)\left(\frac{V\rho v}{\rho_1}\right)\left(\frac{H}{100\,C_p}\right) \qquad (10)$$

which is (1) with $F = 0$

Mixtures

We will now take the more complicated case for mixtures. Start with Fig. 6-8A. This shows how a mixture vaporizes under various pressures. P_1 is the pressure at Point C of Fig. 6-5—P_m is Point D of Fig. 6-5.

Fig. 6-8B is an enlargement of the box in Fig. 6-8A. The pressure lines P_1, P_m, P_2 are assumed parallel for the short interval involved.

Fig. 6-9 gives the vapor pressure. Line AC is at bubble point (zero vapor). Line DF is for V cubic feet/of vapor/100 cubic feet of liquid.

These two lines are also assumed parallel for the small interval involved.

On Fig. 6-9 triangles ABC and DEF are similar, i.e., corresponding sides are proportional.

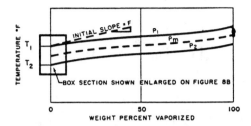

Fig. 6-8A. These Curves Show How a Mixture Vaporizes Under Various Pressures

Fig. 6-8B. *(right)* **Box from Fig. 6-8A Enlarged**

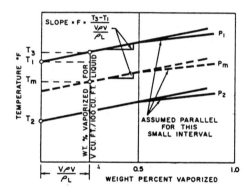

Fig. 6-9. Vapor Pressure of Mixture with Vaporization *(right)*

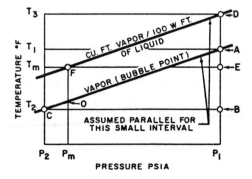

Therefore

$$\frac{P_1 - P_2}{P_1 - P_m} = \frac{T_1 - T_2}{T_3 - T_m} \qquad (11)$$

This can be written

$$\frac{P_1 - P_2}{T_1 - T_2} = \frac{P_1 - P_m}{(T_3 - T_1) + (T_1 - T_m)} \qquad (12)$$

As for a pure substance

$$V\rho vH = 100\rho_1 (T_1 - T_m)Cp \qquad (5)$$

and

$$T_1 - T_m = \frac{v\rho vH}{100\rho_1 Cp} \qquad (6)$$

Also from Fig. 6-8B

$$F = \frac{T_3 - T_1}{V\rho v/\rho_1} \qquad (13)$$

$$\text{or } (T_3 - T_1) = F\frac{V\rho v}{\rho_1} \qquad (14)$$

Substituting (14) and (6) in (12)

$$\frac{P_1 - P_2}{T_1 - T_2} = \frac{P_1 - P_m}{F\dfrac{V\rho v}{\rho_1} + \dfrac{v\rho vH}{100\rho_1 Cp}} \qquad (15)$$

and as before

$$P_1 - P_m = \frac{\Delta h\rho_1}{144} \qquad (8a)$$

Substituting 6-8A in (15) and rearranging

$$h = \left(\frac{144}{\rho_1}\right)\left(\frac{P_1 - P_2}{T_1 - T_2}\right)\left(\frac{V\rho v}{\rho_1}\right)\left(F + \frac{H}{100Cp}\right) \qquad (1)$$

PUMPING LIQUIDS CONTAINING
DISSOLVED GASES

This problem concerns engineers and operators of the various types of petroleum plants *directly* in the following:
(a) Pumping from atmospheric tanks, tanker, barges, tank trucks, etc.
(b) Pumping from tanks that are pressured with gas of some sort.
(c) Pumping from bottoms of steam strippers.
The subject also concerns them indirectly because it is involved in the NPSH tests used at the pump factory.

Air dissolves in hydrocarbons to a much higher degree than it does in water. We do not hear of this making trouble in the refinery. True, many products go to atmospheric tanks. But these tanks are filled and emptied from the bottom. This prevents contamination by air to a great extent. But take other situations. In airplanes, the fuel has to be exposed to the air, and the liquid is certainly sloshed around so that good contact with air is provided. As the airplane attains altitude, what may have been partial saturation at ground level becomes complete saturation at high altitude. In tankers, the liquid is certainly exposed to air, and the rolling of the ship gives good mixing and contact.

Such cases have caused real headaches in pumping. Let us take a case where 100 octane gasoline was pumped. Brown and Bower [1] found that such fluid had 4.5 to 7.0 psi higher true vapor pressure than obtained by what is called the Reid method. Curve D of Fig. 6-10 is a replot of one of their curves.

Some information is missing, but we can piece together a fair picture. Certainly we can be sure from the picture of the test set-up that the gasoline was exposed to air. We can also be reasonably sure that it did not contain methane, ethane or propane. The process of making aviation fuel eliminates these—and probably most of the butane.

Furthermore, if some of these were not removed, they would show up in the Reid type vapor pressure test that was made. In order

[1] Bower, Clark D. and Brown, Perry H., "Prediction of Suction Limitations When Pumping Volatile Liquids," Joint Meeting Southern California Sections of Natural Gas Association and ASME, April 1, 1943.

to make some kind of comparison, I computed Curve C_1. To avoid extremely difficult complications, I had to assume a pure single component having a vapor pressure of 6 psia at 73.5°F. I assumed saturation with air at 14.696 psia and 73.5°F. The solubility of nitrogen was obtained from NGAA equilibrium curves. Oxygen is guessed at— guided by data for water.

Curve C_2 is a reference curve for the hydrocarbon of Curve C_1, but absolutely free of air. While my method of comparison is not too scientific, it does bring out for sure that the sample of Curve D was not fully saturated with air.

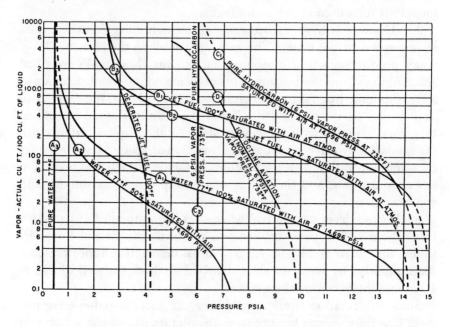

Fig. 6-10. Equilibrium Vaporization of Water Liquids

It might be well to take an example of what might happen to a design engineer. The pump manufacturer might quote a pump for 6 feet NPSH to pump the gasoline of Curve D. And whoever furnished the gasoline would follow the custom of the industry and would tell him that the vapor pressure at pumping temperature was

6 psia. So the engineer would take 6 feet and divide by 3.0 ft/psi (for gasoline) to get 2 psi. This, added to 6.0, gives 8.0 psia that would be needed at the pump nozzle to have it work properly. In other words, he would allow 6 feet of liquid above the supposed vapor pressure at the pump intake.

But suppose that the gasoline had some air dissolved. If partially saturated as per Curve D, 10 cubic feet of vapor would be formed per 100 cubic feet of liquid—and it still hasn't reached the eye of the impeller.

It could be worse. If thoroughly saturated by transportation per example, it might follow something like Curve C_1. Then, over 100 cubic feet of vapor would form per 100 cubic feet of liquid at 8 psia at pump intake. Certainly, someone would be in trouble.

An actual case[2] of pumping absorption oil falls right into a similar situation. The pump used required 6 feet NPSH by open cold water test at 80 gpm. On abosrption oil, this meant that the pump should work with 2.35 psia at the suction nozzle. Yet at 11.5 psia, it reached cavitation.

Another way of looking at it, 30 feet NPSH was required with absorption oil as against 6.0 feet for cold water. It is almost a certainty that dissolved air was the cause of this discrepancy. The absorption oil was transported to the test site by truck. It was loaded and unloaded. Both these actions increase air contact. And the oil was exposed to air during test.

Some interesting data[3] on jet fuel are given on Curves B_1, B_2 and B_3 of Fig. 6-10. These curves represent true vapor pressure taken in the laboratory. B_1 and B_2 are for samples thoroughly saturated with air. It looks as though the sample used in Curve B_1 was saturated at a lower temperature than $100°F$ because its bubble point is higher than atmospheric pressure.

The sample for B_3 was thoroughly deaerated before it was placed in the true vapor pressure device. While these curves hardly fit into

[2] Hendrix, Lloyd T., "The Determination of Hydrocarbon NPSH Characteristics," California Natural Gasoline Association, October 10, 1957.

[3] Personal Communication with Aerojet Corp.

refinery problems, they are quite valuable information. We know for certain the degree of saturation.

There was another interesting case of apparent high NPSH where a pump handled an amine solution. The solution contained carbon dioxide in chemical combination. However, the carbon dioxide could be liberated into solution if given time and some degree of temperature.

The solution flowed from the regenerator through exchangers and was pumped before final cooling. This provided time and a fair degree of temperature. The pump required an apparent NPSH of some 30 feet, whereas an open cold water test showed some 11 feet. The solution to the problem was a slow speed booster ahead of the main pump.

The booster requires very little NPSH. It may have had cavitation, but it did not register. The main pump ruined itself in a few days since it had a 200-hp motor. The booster had a 15-hp motor. Without doubt the damage done is related to the power.

Curve A_1 is computed for water, assuming 100 percent saturation at atmospheric pressure. Henry's law constant was used for nitrogen and oxygen. Curve A_2 assumes 50 percent saturation. Unfortunately, we run into a real snag when we try to apply these curves to cold water tests. We don't know the degree of saturation, and we don't know how much vapor the impeller will handle.

We can speculate a little, however. We know that we get apparent NPSH's of 6 to 12 feet in certain pumps during open cold water tests. This means (for 77°F) that the pressure in the suction nozzle was 3 to 6 psia. But what degree of saturation do we use? We might try 50 percent. Then we would have 0.55 to 3.5 cubic feet of vapor per 100 cubic feet of liquid *at the pump nozzle*.

There is still some pressure drop required to get flow into the impeller. So we have another big guess. It is one guess after another. With our present state of knowledge, we cannot get answers.

It takes a different approach. A good one would be to use a setup using boiling water. Positive boiling, both before and during the test, would eliminate air. NPSH is best measured by varying the height in the suction tank. This would not tell us how much vapor formed in the impeller. But the measured NPSH would be very close to the absolute NPSH.

If we look at Fig. 6-6B, we see that .008 foot suppression gives 2 cubic feet per 100 with water at 14.7 psia (212°F). Thus, 20 cubic feet per 100 would require 0.08 foot, and 80 cubic feet per 100 would require 0.32 foot. We should certainly get very close to the absolute value of the NPSH.

Of course, we are not going to get all this nice data very soon. And even if we had them, we still do not always know what we are pumping. In the meantime, we have to design pump installations. Each of us will probably work out his own set of general rules and apply judgment in special cases. We might write down this set of suggested rules:

- Take open cold water NPSH tests as absolute. True, they may contain safety factors, but maybe less than we think.

- In pumping from refinery atmospheric storage where filling and withdrawal is from the bottom, use a vapor pressure halfway between 14.7 and Reid type.

- In pumping oil from a steam stripper, assume oil is at boiling point. Remember that steam dissolves in oil too.

- In pumping from tankers, tank trucks, etc., where contact with air is accelerated, assume saturation. Specially designed or selected pumps or boosters may be needed.

7

Bypasses and Partial Flow in Centrifugal Pump Operation

Centrifugal pumps are used under widely varying and often adverse operating conditions. When uninterrupted pump operation is required, pump failure causing a single day of lost production may represent a significant revenue loss. Hence, an effective analysis of the centrifugal pump operation is important to plant engineers.

Many engineers overemphasize full-flow pump efficiency at the best efficiency point. Single-point performance evaluation ignores partial-flow performance, particularly when high-energy pumps are used in parallel operation. It is possible that a design selection based only on the full-flow condition may result in marginal performance at reduced flows. Thus, good peak-efficiency performance at design conditions may not assure good overall operating reliability. Changes in delivery requirements can easily cause the pump to operate in an off-design-flow condition and to require frequent rebuilding. Therefore, a careful look at the anticipated range of normal operation is desirable. Planning for operation of pumps at reduced capacity is also important for reducing the frequency of overhauls.

106

Bad entry, poor bypass design, inadequate net positive suction head available, excessive prerotation, and off peak-efficiency operation contribute to poor pump performance. If these unfavorable factors are not corrected, the pump impeller may quickly be damaged beyond repair.

Cavitation of a centrifugal pump causes mechanical deterioration and, ultimately, failure. Performance curves may continue to indicate that a pump is operating satisfactorily. But even slight deviations in a performance curve can signal a condition that will result in cavitation damage.

An additional and equally restrictive limitation on pump operation is the unbalanced axial loading on the impeller, and also the overheating which may occur if the pump is operated at an excessive distance on the pump curve to the left of the BEP. While both conditions limit the life of the pump, the temperature increase may cause immediate damage, if not destruction of the pump. The problem is aggravated in units which are handling high temperature fluids.

Because of the modern high performance pump requirement for low NPSH, engineers sometimes specify a larger pump than is required for the estimated flow conditions. Inasmuch as the pump will operate, statistically, at full load for only a few percent of the time, operation is most often at the shut off end of the pump curve. However, even with a properly sized pump, operation will frequently be at the low capacity end of the curve.

Furthermore, it is not unknown that a pump will operate against a closed shut-off valve due to operational error. Double suction pumps are particularly susceptible to cavitation when operated at low flows. Low-flow cavitation has been a challenge to pump designers and operators for many years. This type of pump damage is quite different from that which occurs in pumps operating at excessive capacity or with less than rated NPSH.

An obvious and common solution to minimize the effects of low flow is the provision for bypassing a certain amount of the pump discharge back to the suction source. The prime criterion for determining the amount of bypass flow is almost always the temperature rise in the pump. Several methods may be utilized: a fixed orifice, calculated to bypass enough discharge to limit temperature rise in the pump; an orifice automatically modulated to provide the necessary bypass; solenoid controlled bleed-off which opens when a flow sensor determines that bleed-off is required; a modulating bleed-off valve controlled by a flow sensor and an external controller.

The question then is: what is the necessary amount of bypass required to safeguard the pump? There are two possible approaches to the answer—depending on whether the pump is a relatively low-head unit, or whether it is a boiler feed pump with gland bleed off. For many years, a rule in widespread use was to use 30 gpm/100 hp at shutoff. This rule was convenient, as it was safe and reasonably economical. The rule was chosen to limit the temperature rise across the pump to 15°F. A simple formula corresponding to the 30 gpm rule, but allowing for a choice of temperature rises, is

$$E_p = \frac{TDH \times 100}{778 \times t + TDH} \tag{1}$$

in which

E_p = Pump Efficiency, expressed as a percentage;

TDH = the Total Dynamic Head;

t = Allowable Temperature Rise, expressed in deg. F.

The capacity corresponding to this efficiency is found on the pump characteristic curves.

Both of these methods are satisfactory for low performance pumps. It has become apparent, that as unit sizes and operating pressures have increased, these simple methods would require inordinately large recirculation flows. More precise analysis will pay off.

Pumped liquid temperature rise, when based on published formulae, assumes that all liquid within the pump is raised to the same temperature level. Observations of operating pumps indicate that this assumption is in error. It has been determined that a number of losses affect the temperature rise in the pump itself. These are, generally, hydraulic, bearing and stuffing box, disc friction, leakage, and backflow circulation.

A rigorous method for determining required bypass flow is contained in *Hydrocarbon Processing* for July 1965, Vol. 44, No. 7. This method may be useful in sophisticated installations, or where exotic fluids may be used. A somewhat simpler method is contained in *Combustion* for July 1976, as follows:

"Most modern barrel-type boiler feed pumps are so constructed that neither the materials used nor the internal clearances are apt to be the limiting factors insofar as the temperature rise across the pump is concerned. Where O-rings are used internally, these are made of EPR (ethylene propylene rubber) and can work continuously at temperatures as high as 450°F. Since most boiler feed pumps have suction temperatures of 300 to 400°F, material selections would permit temperature rises approaching 50 to 150°F, depending on the suction temperature. In other words, temperature rise limitations are essentially dictated by the need to prevent flashing within the pump.

"Modern analysis of temperature rise through pumps shows that all the brake-horsepower applied to the pump shaft does in fact increase the enthalpy of the feedwater passing through the pump. The point of highest temperature occurs in the balancing device leak-off chamber and it is this area which must be protected against flashing when determining the recommended minimum flow. (See Fig. 7-1.)

"While a more rigorous method is available, it is most convenient and safe to use the formula:

$$\Delta T = \frac{\text{Total Head in feet}}{\text{Efficiency} \times 778} \qquad (2)$$

where

ΔT = Total temperature rise across the pump and balancing device.

"To illustrate how this works, we will assume the following additional data for your projected installation:

Suction pressure	=	140 psia
Pump curve	—	as shown on Fig. 7-2
System-head curve	—	as shown on Fig. 7-3

"Assuming variable speed pumps, the usual practice is to assume that the maximum speed and head that a pump coming on the line will be required to operate at will be that required when one pump is on the line (A) at the one-pump run-out conditions and the second pump (B) is just at the speed and head needed to lift its check valve in order to pump into the system. In your case, with the assumptions we have made, this would occur at a speed of 4700 rpm (basis 5600 rpm full speed) and a head of 6300 feet.

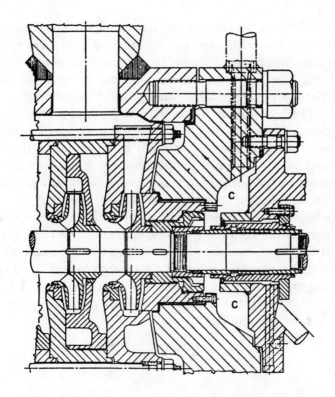

**Fig. 7-1. Typical Arrangement of Boiler Feed Pump Balancing
Device Showing Balancing Leak Off Chamber "C"**
(Source: *Combustion* July 1976)

"Having determined the maximum head at which the pump in
question may be operating at the moment when it requires protection against overheating, we must now determine how much temperature rise the feedwater passing through the pump should be allowed
to experience. This maximum temperature rise cannot be calculated
from a rigid mathematical formula, but rather on some empirical
basis which will allow for a contingency margin. Most installations
made today provide for the return of the balancing leak-off back to
the suction side of the casing. Empirical considerations dictate that
the vapor pressure of the feedwater within the balancing device

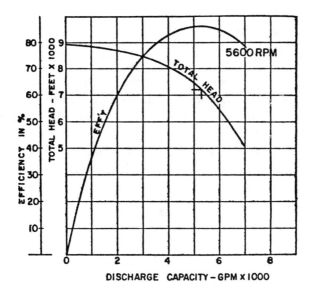

Fig. 7-2. Boiler Feed Pump Curve

(Source: *Combustion*, July 1976)

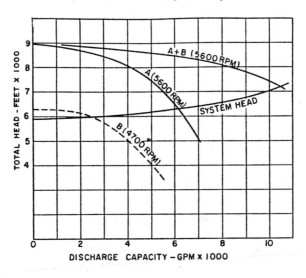

Fig. 7-3. System Head Curve

(Source: *Combustion*, July 1976)

leak-off chamber not be permitted to exceed a value equivalent to the pump suction pressure less 10 psi:

Vapor Pressure in balancing leak-off chamber $\leq P_{suction} - 10$ psi

"In your example, this maximum vapor pressure would be (140–10) or 130 psia, which corresponds to a maximum temperature of 347.3°F. The maximum allowable temperature rise is then 347.3–316=31.3°F which most people would round off to 30°F.

"Since the total temperature rise is a function of head and efficiency, the minimum permissible efficiency can be computed as follows:

$$\text{Eff. min.} = \frac{\text{Total Head}}{\text{Max. }\Delta T \times 778} = \frac{6300}{30 \times 778} = 0.27 \tag{3}$$

or 27 percent.

"On the pump curve we have assumed (Fig. 7-2), this corresponds to a capacity of 750 gpm at 5600 rpm. It should be noted, however, that the head produced at this speed is 8900 feet, whereas the accelerating pump will lift its check valve once it produces a total head of 6300 feet (or less). Now the speed at which the pump will develop 6300 feet and have an efficiency of 27 percent can be calculated to be

$$5600 \times \frac{6300}{8900} = 4710 \text{ rpm} \tag{4}$$

"The required minimum flow will be:

$$750 \times \frac{4710}{5600} = 618 \text{ gpm} \tag{5}$$

"Starting with the assumptions we have made, it appears to us that the minimum flow figures you have been given were probably computed on the basis of an allowable safe temperature rise calculated by the method we have illustrated. While we might instinctively prefer a more conservative approach, it is clear that the earlier empirical rule of 30 gpm/100 bhp at shut-off may have been excessively conservative in certain cases. Where conditions prevailing in the balancing device leak-off chamber are the governing consideration—as it usually is in the case of fossil units—the method we have outlined yields an easily determined and a safe value.

"Some precautions do merit considerations. For instance, we recommend that the suction temperature used in determining the maximum temperature rise be that at full load, even though pump A in our illustration could only carry 60 percent of the load when running alone. On hot restart situations, it is possible that pump B could be coming back on the line before the suction temperature had had enough time to decrease significantly. The use of the equilibrium temperature corresponding to one-pump run-out conditions could result in an inadequately sized recirculation system.

"A second important precaution is to design the recirculation system for the maximum head that the idle pump could develop. This seems clear enough in the case of two half-capacity turbine driven pumps, but is not as obvious in the case of, say, a third half-capacity motor driven pump operating at variable speed by means of a hydraulic coupling. It would be possible to operate this pump at the design total head of 7200 feet while the two regular half-capacity pumps were in operation. The designer of the recirculation system must know whether he needs to allow for this contingency when setting the minimum allowable flow using the allowable temperature rise approach we have described.

"Finally it should be noted that most recirculation systems are designed on the assumption that operation at minimum flow will be of relatively short duration, probably not exceeding 100 to 150 hours per year. Where this is not the case, accelerated impeller wear will occur, which will vary with the nature of the application and of the pump involved. We have even known cases where operation of a pump at these minimum flows was impossible because of hydraulic vibrations and where the minimum flow had to be increased to a value where vibrations disappeared.

"With some exceptions, considerations of this nature are almost exclusively characteristic of nuclear power plants, which operate at very light loads for considerable periods before final authorization for full load operation is obtained.

"However, we do want to highlight the fact that in any plant, be it fossil or nuclear, unusual circumstances may require extended operation of the pumps at or near the minimum flow. If this is the case, the fact must be made known to the manufacturer who will then be guided by other than thermal considerations in recommending a minimum flow system.

"One more word of caution: we have known of installations where the time of response to low flow conditions is excessively long and where a pump may be made to operate against a completely closed discharge for too long a time. The better part of valor is to provide such interlocks that the recirculation valve of a pump about to be brought on the line cannot be in a closed position when the pump is started."

CAVITATION AND MINIMUM FLOW PARAMETERS FOR IMPROVED RELIABILITY OF FEEDWATER PUMPING EQUIPMENT*

Cavitation Parameters

Over the years, increasing power-plant sizes and physical equipment placement limitations (for instance, limitations on DA heater height) have forced continuous efforts by pump manufacturers to reduce Net Positive Suction Head (NPSH) requirements on feedwater cycle pumps. One innovation developed to meet this need has been the so-called large-eye impeller that requires less NPSH at rated and run-out flow conditions. (See Fig. 7-4.)

However, this improved NPSH performance was not without some undesirable trade-offs. One of the first symptoms was abnormal noise at low flows. Increased vibration was also noticed. Then a form of impeller cavitation damage occurred early in equipment life. The one common denominator was that all these symptoms occurred during low-flow operation where the margin between NPSH available and NPSH required was thought to be high.

Another trade-off for low NPSH impellers can be low-frequency hydraulic pulsation in the range of about one to eight Hz. This low-

---

*Abstracted from ASME 79-JPGC-Pwr-8, "Design and Application of Feed-Water Pumping Equipment for Improved Availability in Cyclic Operation," by W. S. Eadie, Pacific Pump Division, Dresser Industries.

Fig. 7-4. Cut Away View
Damage to the back side of the first-stage impeller results from continuous operation at low flow. It is caused by flow recirculation and
poor vane angle/flow match at these low flows.
(Source: ASME 79-JPGC-Pwr-8)

frequency pulsation is produced by impeller-eye recirculation, and
may be detected at low flows less than 60 percent of the best efficiency point. Since low-flow recirculation is greater in large-eye impellers, this particular type of pulsation is generally associated with
these designs.

Pump impeller damage resulting from this problem looks like cavitation damage, but there are several significant differences between
it and classic cavitation due to inadequate NPSH. It occurs on the
back side of the impeller vane and tends to be concentrated near
the periphery of the impeller inlet eye, Fig. 7-4. At low flows, inlet
recirculation promotes damage at the periphery of the impeller eye.
Impeller vane inlet angles, determined on the basis of design flows,
do not match inlet fluid angles at low flows. This produces areas of
very low pressure on the back of the impeller vanes which results in
localized flashing and cavitation.

Unfortunately, the actual capacity at which this form of cavitation begins to occur cannot be readily determined on test. Unlike cavitation caused by inadequate NPSH available, which can be detected by loss of head on the test stand, this phenomenon is not readily detectable by head loss. Any performance change occurs only after considerable damage has been done.

Thus, there is a relatively poor trade-off here. Large-eye impellers will improve NPSH characteristics at best efficiency and beyond, *but* this is accomplished only at the expense of probable impeller damage at lower flow as shown in Fig. 7-4.

A comparison of NPSH performance for a multistage boiler feed pump with large- and normal-eye impellers is shown in Fig. 7-5. Note that at about 60 percent of the flow at best efficiency, the NPSH required curves cross. Therefore, at flows below this 60-percent point, the large-eye impeller requires *more* NPSH and, as has been described, is more susceptible to back-of-the-vane cavitation damage at low flows. (Since double-suction impellers can also be designed with large eyes, they too suffer the same characteristics and the potentially damaging effects.) Large minimum flow values are costly in power consumption, and therefore, some consideration should be given to this trade-off problem when pump operating conditions are specified.

The minimum flow based on impeller damage criteria from impeller-eye recirculation can be lowered if Suction Specific Speeds (S) are limited to 8500 or lower at the best efficiency point, and available NPSH is set realistically, taking into account adequate margins for system upsets and the severe conditions inherent in cyclic operation. A 50-percent (or more) margin of NPSH available over NPSH required at the best efficiency point, even though it may mean raising the DA heater or increasing booster-pump pressure, will result in reduced maintenance costs and improved pump availability.

Thus, the minimum NPSH that should be made available by power-plant designers at the point of best efficiency is shown in Fig. 7-5A.

Shop performance testing of any feed-cycle pump set should determine the actual NPSH required throughout the operating range (from minimum flow to the intersection with the system head curve).

In vertical, multistage feed-cycle pumps, such as condensate and

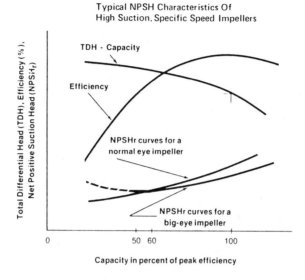

Fig. 7-5. NPSH required performance of a high suction specific speed impeller is better at rated flow and beyond. But at less than about 60-percent capacity, NPSH required for these impellers rises rapidly.   (Source: ASME 79-JPGC-Pwr-8)

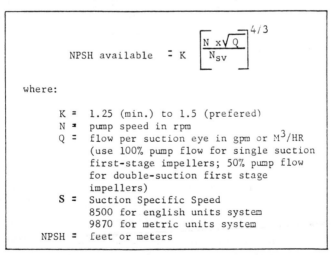

Fig. 7-5A (Source: ASME 79-JPGC-Pwr-8)

heater-drain pumps, increased pump flows resulting from larger units have presented new problems. The higher NPSH required on larger units have made it necessary to elongate the pumps well beyond the minimum required length to accommodate the necessary number of pump stages. This means deeper pit excavation, longer pump shafts, and more intermediate bearings. Reducing pump NPSH required through reduced pump speed is counterproductive. The added stages needed can more than offset the NPSH reduction and the cost penalty is severe.

One solution to this problem is the use of an axial-flow first-stage impeller, to reduce NPSH required over the entire operating range. (Fig. 7-6)

Successful operation of axial flow impellers at suction specific speeds to 19,000 can be expected. In a typical multistage condensate pump requiring say 22 ft (6.7mt) NPSH at the impeller, the use of an axial-flow, first-stage impeller can cut this NPSH about 14 ft. It should be a totally separate pumping element with straightening vanes between it and the second-stage impeller inlet to preclude prerotation. The axial-flow impeller's multiple vanes insure hydraulic stability.

Vertical, multistage pumps in feed-cycle service should always be shop tested at full assembled length and rated operating range. As is the case with all other feed-cycle pumping equipment, careful attention should be paid to minimum flow quantities based on the total system considerations.

Minimum Flow

For many years by-pass systems have been designed for the sole purpose of protecting boiler feed pumps against the disastrous effects of rapid temperature rise at minimum-flow condition. Based on temperature rise considerations alone, minimum pump flow has been considered to be as low as 10 percent of rated pump flow. Flow control consisted of a fixed pressure-breakdown orifice in series with a suitable valve. However, problems have developed indicating that temperature rise protection is not the only criterion. While

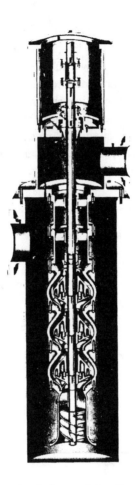

**Fig. 7-6. Axial-Flow, First-Stage Impellers Require Less NPSH
and Thereby Shortens Required Pump Length**

about 10-percent minimum flow may be satisfactory in some cases
from a temperature-rise standpoint, it is not sufficient to prevent
long-term impeller erosion problems resulting from the back-of-the-
vane cavitation phenomenon previously discussed. This problem has
become particularly pronounced since cycle economics began to
mandate high-speed boiler feed pumps.

Experience has shown that back-of-the-vane cavitation phenomenon can drastically reduce impeller life, if long-term operation is permitted at flows significantly below the pump's capacity at best efficiency. This effect is particularly noticeable on impellers with very low NPSH characteristics (high suction specific speed).

Based on empirical evidence it appears that for a typical suction impeller with low suction specific speed, the relationship of impeller life to minimum flow (percentagewise and based on continuous operation) is roughly as represented by Fig. 7-7.

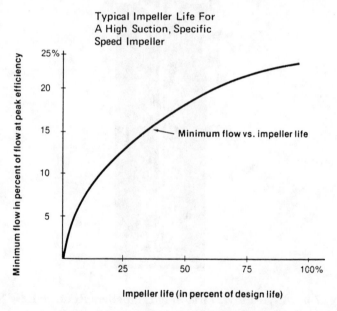

Typical Impeller Life For
A High Suction, Specific
Speed Impeller

Minimum flow vs. impeller life

Minimum flow in percent of flow at peak efficiency (y-axis)

Impeller life (in percent of design life)

Fig. 7-7. Long-Term Operation at Flows Significantly Below the Best Efficiency Point Causes Severe First-Stage Damage in Feed Cycle Pumps
(Source: ASME 79-JPGC-Pwr-8)

But since experience indicates that the percent of B.E.P. at which the effects of recirculation are minimized varies with pump size and S, we have attempted in Fig. 7-7A to graph recent experiences concerning impeller life in actual power-plant operation. While further

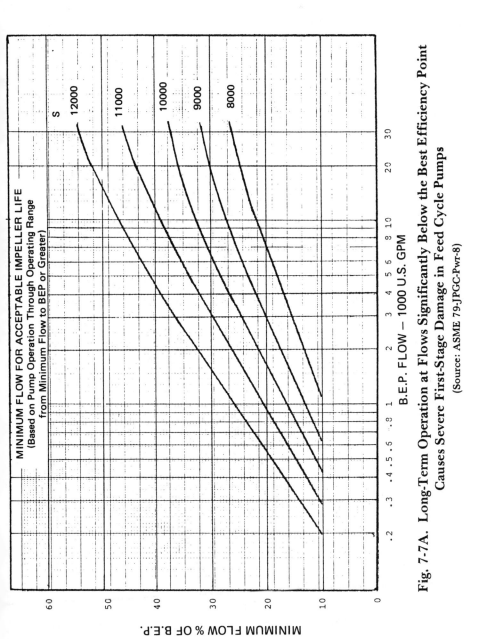

Fig. 7-7A. Long-Term Operation at Flows Significantly Below the Best Efficiency Point
Causes Severe First-Stage Damage in Feed Cycle Pumps

(Source: ASME 79-JPGC-Pwr-8)

experience may refine this approach, we submit this chart as a useful guide for making the trade-off between minimum-flow requirements (based on impeller life) and the capital costs necessary to provide the additional NPSH. Substituting the S value of chart 7-7A in place of 8500 in the formula (Fig. 7-5A) will allow determination of NPSH requirements.

Minimum Flow Control

It should be pointed out that in the case of a large boiler feed pump, say in the 15,000-hp range, the economic loss based on a 25-percent minimum flow is very high. Therefore, a fixed-orifice by-pass system cannot be tolerated. A modulating system is far more economic since it limits actual by-pass flow to only that amount actually needed taking into account boiler demand under any load condition.

Some of the reasons why modulating bypass, minimum-flow systems have become predominant on modern boiler feed pumps are:

— Modulating by-pass valves combine the functions of on/off valves and the minimum-flow control office.

— Reliable valves are now available. They operate at low noise levels and provide extended service life even at supercritical pressures.

— Their smooth control eliminates hydraulic shock to the system.

— They conserve fuel and operating costs because only the incrementally required flow is by-passed. Fig. 7-8 (a) and (b) shows a by-pass flow comparison between on/off systems and modulating systems. And for example, in a 500-MW plant, the excess by-pass flow for an on/off system can cost over $500 per hour.

— The so-called deadband inherent in any on/off system to prevent violent valve hunting around the set minimum-flow value is eliminated.

In selecting modulating control valves, they should preferably be designed to be readily disassembled and easily cleaned since plugging

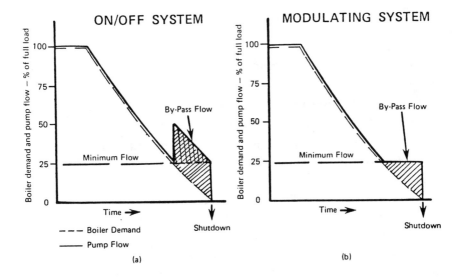

Fig. 7-8. This comparison of boiler feed pump by-pass flows during shutdown in on/off and modulating systems shows the excess by-pass flow required (a) to prevent valve hunting in on/off systems. In a modulating system (b), this costly excess flow is eliminated.

(Source: ASME 79-JPGC-Pwr-8)

can occur within the orifice portion of some valves. Also, a long stroke from fully open to fully closed promotes close modulation. Since the operating pressure across these valves is very high, seating arrangements which will eliminate costly leaking in operation are to be preferred.

Modulating valves should be individually shop tested to determine their actual performance. And this is best done with boiler feed pumps.

Warm-Up

Adequate equipment warm-up provisions for start-up purposes and for the maintenance of operating temperatures in standby equipment serve two purposes:

1. It permits pumps and drivers to approach their operating alignment prior to start-up.

2. It eliminates possible internal misalignment that can result from unequal thermal expansion between relatively thick and thin sections within the pump. The possibility of distortion and internal rubbing is high if the pump is not properly warmed up and maintained in its fully warmed-up condition prior to start-up.

Full insulation—beneath as well as over the top of the pump—minimizes temperature stratification within the pump and turbine driver and reduces the amount of warm-up flow required. Temperature detectors should be permanently mounted at a minimum of four places (top and bottom at both ends) on boiler feed pump casings to monitor temperatures throughout the pump. Seal injection flow controls should be set so that cold-injection flows only towards the drain end of the seals in idle or stand-by pumps to prevent cool-down of that portion of the pump.

In general, all parts of a pump should be held within 50°F (10°C) of each other and to within 75°F (23.9°C) of pumping temperatures. If there is one thing worse than bringing a cold pump on the line, it is bringing a partially warmed-up pump on the line. Under this latter condition, internal misalignment is almost certain to exist, and the possibility of pump damage is high.

GENERAL BYPASS CONTROL*

All factors of a pump bypass system must be considered to assure positive pump protection at low flows. These factors include: capacity; positive, reliable operation; long life; quiet operation; power requirements; system simplicity; system design time; pressure breakdown ability; and installed cost.

To ensure proper sizing of a centrifugal-pump bypass system, regardless of the fluid being pumped, it is best to rely on the specific pump curve and recommendations of the manufacturer. As a minimum, this information should be provided to the prospective supplier of a bypass system: (1) maximum pump capacity; (2) bypass flow quantity; (3) pump shutoff head; (4) sump or reservoir pressure; (5) fluid to be pumped; and (6) fluid temperature and specific gravity.

Three basic types of recirculation systems are used in industry today: continuous recirculation, flow-controlled recirculation, and automatic recirculation control.

Continuous Recirculation Systems

In a continuous recirculation system, the pump and its prime mover are oversized to provide enough fluid to keep the pump cool by continuous recirculation from the discharge of the pump, back to the sump (or reservoir) on the suction side.

A fixed orifice in the recirculation line breaks down the pressure differential between the pump discharge and the reservoir. This orifice is sized to continuously recirculate sufficient fluid to keep the pump cool. But because of this continuous recirculation, this fluid never adds to the value of the final product. The system is, therefore, inefficient and very costly because:

- Both the pump and its prime mover must be oversized to handle the recirculated quantity even at full load on the pump, where there is sufficient flow to prevent overheating of the pump.

- The additional power costs to run the pump driver are higher than normally realized. For example, a pump in a 0.9¢/kWh area discharging 300 gpm at a 2,500-ft head, requires 50 gpm recirculated flow to keep it cool (Fig. 7-9). Assuming the water to be at 330°F, extra annual power costs may be calculated as follows:

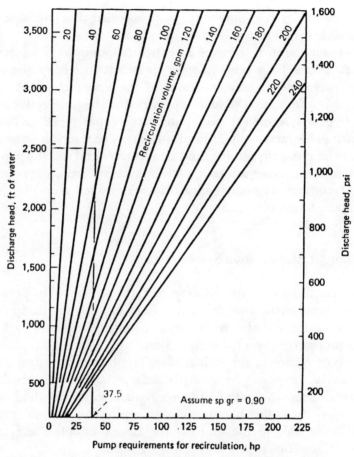

Fig. 7-9. Horsepower Required for Pump Recirculation
(Source: *Chemical Engineering*, Feb. 4, 1974)

$$hp = \frac{(gpm)(\text{discharge pressure, ft})(\text{sp gr})}{3,960 \ (\text{pump efficiency})} \qquad (6)$$

$$= [\,(50)(6,535)(\text{power cost})\,]\,/\,[\,(3,690)(0.75)\,]$$

$$= 37.5 \text{ hp over that needed for process demands}$$
(see Fig. 7-9)

$$\text{Annual cost} = [\,(hp)(6,535)(0.9)\,]\,/\,(\text{driver efficiency})$$

$$= [\,(37.5)(6,535)(0.9)\,]\,/\,0.85 = \$2,600$$

Thus, \$2,600 would be the cost to recirculate sufficient fluid to cool the pump (Fig. 7-10). This cost would gradually increase as the fixed orifice began to wear and enlarge due to the high fluid velocity through it, which would then cause an increase in the quantity of fluid passing through it. This, plus the original purchase costs for over-sized pumps and drivers, make a continuous-recirculation system prohibitively expensive except on the smallest of pumps.

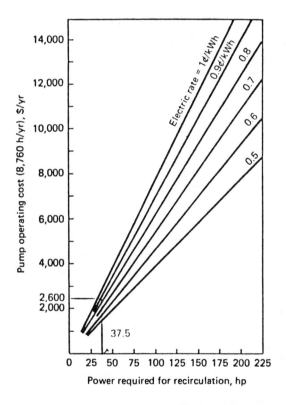

Fig. 7-10. Annual Cost of Recirculation (Continuous Flow)
(Source: *Chemical Engineering*, Feb. 4, 1974)

Flow-Controlled Recirculation

In a flow-controlled system (Fig. 7-11(a)), a fluid is recirculated only when flow through the pump approaches the minimum specified by the pump manufacturer.

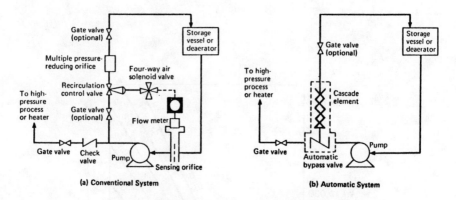

Fig. 7-11. Recirculation-Control Systems
Conventional Arrangement is Shown in (a), the Automatic Kind in (b)
(Source: *Chemical Engineering*, Feb. 4, 1974)

As the orifice on the suction (or discharge) side of the pump senses that flow through the unit nears the minimum requirement, the flow transmitter sends an electrical signal to open the solenoid valve, thus opening the recirculation-control valve. Fluid is now recirculated to the storage vessel on the suction side of the pump in a quantity governed by the size of the recirculation-control valve and the pressure drop across it. Fluid may also be going to the process.

A multiple pressure-reducing orifice, located after the recirculation-control valve, reduces the pressure drop from the discharge of the pump to the reservoir. This pressure drop can be very large at high pump-discharge pressures. At the instant of opening and the instant of closing, the recirculation-control valve receives the entire pressure drop from the pump discharge to the reservoir, and because of this is subject to severe wire-drawing and erosion.

The control valve must also shut tightly against full pump-discharge pressure. Any leakage through the valve should be subtracted from the total pump or system capacity. At times of severe leakage, the pump may not be able to satisfy the system at full-flow requirements.

A recent survey of electric utilities indicated that the single most troublesome control valve in their system was the recirculation-control valve. Repair is frequently due to the high pressure-drop, and high inlet or shutoff pressure. In any type of recirculation system, the recirculation-control valve should fail open to completely assure protection of the pump, in the event of air or electrical system failure.

Variations of the flow-controlled system include a temperature-differential system wherein the temperature rise across the pump is used to open and close the recirculation-control valve. Thermocouples are used to measure suction and discharge temperatures. There have been difficulties, however, in determining the proper placement of sensing elements.

When electric motors are used to drive pumps, it is possible to control the recirculation system by measuring motor amperage, which is a function of the work, and therefore of the flow through the pump. But neither of these latter systems are in frequent use today.

All the systems so far discussed include a check valve on the discharge side of the pump to prevent reverse flow through the pump. Flow- and temperature-controlled systems also require both electric power and a pneumatic system to actuate the control valves.

Automatic Recirculation Control

Automatic-recirculation control can be provided in a system that combines the functions of the pump discharge check valve—and the recirculation and pressure breakdown elements—in a single, self-powered unit (as shown in Fig. 7-11(b).

The described system is that manufactured by Yarway, although the ConTec Company, and possibly others, manufacture bypass controls to operate similarly.

In this system, the rising disk-type of check or nonreturn valve acts as a flow-sensing element. This opens and closes a small pilot valve, which triggers the opening and closing of the recirculation-control valve. This is a tight shutoff element, whose unique cascade design dissipates the high-pressure energy prior to passing the recirculated fluid back to the low-pressure sump or reservoir.

With normal main flow (Fig. 7-12(a)), the check valve rises off its seat and floats on the load or flow discharging from the pump. The lower extension of the check valve lifts the left end of the lever arm, thus allowing the pilot valve to seat to prevent flow through the lower passage to the low-pressure system.

Full pump-discharge pressure on the head of the pilot valve keeps it well seated. This pressure is also exerted on the head of the cascade piston—via the annular clearance around the pilot-valve stem—which closes the cascade valve and prevents bypass flow. Pump pressure is also applied on the opposite side of the piston, but to a smaller area. Thus, the cascade recirculation valve remains closed.

As process requirements decrease and pump capacity is reduced, the spring-loaded check valve begins to descend toward its seat (Fig. 7-12(b)). The lever arm, through its pivot point, now can open the pilot valve. When the pilot valve opens, the high pressure on the head of the cascade-valve piston is vented downstream to the low-pressure bypass portion of the system. As this happens, the piston moves to the right—due to pressure imbalance—and recirculation flow begins.

The point at which the pilot valve opens to place the system in the bypass mode is carefully calculated to match individual pump characteristics. It is controlled by the annular clearances between the tapered lower portion of the check-valve disk and the surrounding body. The check-valve disk becomes, in effect, a variable-area flow-meter when pump-discharge quantities fall below 40 percent of pump capacity.

When the check valve is fully seated due to lack of process-flow requirements (as shown in Fig. 7-12(c)), the cascade piston and valve are fully open. Bypass flow, which then is at its maximum, is returned to the low-pressure sump or reservoir.

Although the tightly seated check valve prevents reverse flow

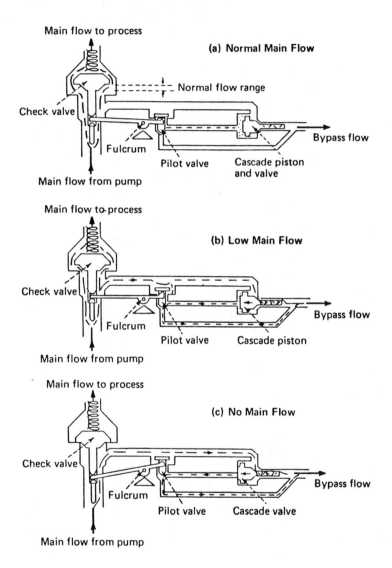

Fig. 7-12. Bypass-Control Valve Mode of Operation
(Source: *Chemical Engineering,* Feb. 4, 1974)

through the pump when the pump is out of service, reverse flow through the bypass system can be used to keep the pump casing and internals warm when the pump is not being used.

The cascade valve controls the bypass flow and dissipates the high-pressure energy differential from the pump discharge to the sump. This valve splits the main velocity flow into multiple streams.

Parallel flutes, milled into the cascade cylinder, direct fluid flow through a series of 90-degree turns as it cascades through the valve. The flutes constitute a series of variable orifices, and each set of flutes (or "stages") absorbs part of the fluid energy. The number of stages is governed by the magnitude of the pressure drop across the valve. The size or depth of the milled flutes governs flow capacity.

Because the seating surfaces are not exposed to high fluid velocity, they maintain tight shutoff of the bypass system for long periods. And since no external source of power—electrical or pneumatic—is needed, the system is designed to fail-safe, guaranteeing minimum flow through the pump. Virtually any magnitude of pressure drop can be handled by the system.

For less demanding applications—low temperature, moderate pressures, and where protection against valve shut-off is desired—simpler methods may be used. As a matter of fact, there are many installations where bypass provisions are not considered. Many times, this philosophy turns out to be expensive in terms of reduced pump life or possible damage due to excessive temperature rise in the pump.

*Assuming that there are no high pressure gland leak-offs, and neglecting bearing and transmission losses, which are usually negligible, the difference between the brake horsepower of a pump (power input) and the water horsepower (useful work done) is converted into heat, which results in a temperature rise of the liquid being pumped. If the temperature rise is excessive, it can result in thermal expansion of rotating parts and ultimately in seizure, damage to packing, and flashing of the liquid with consequent vapor binding.

*This section is abstracted from *Heating, Piping & Air Conditioning,* January 1966.

The temperature rise can be determined from the formula:

$$\Delta t = (H/778C_p) \, [(1 - e)/e] \tag{7}$$

where

  $\Delta t =$ temperature rise, °F
  $H =$ total dynamic pump head, ft
  $C_p =$ specific heat of liquid pumped, Btu per lb per deg F
  $e =$ pump efficiency, decimal

This temperature rise is usually not too serious for cold liquids, but it should be considered when hot liquids are being pumped and especially when a pump is being operated at a small fraction of its rated output. At low flow rates, pump efficiency is greatly decreased.

To prevent excessive temperature rise, a minimum flow must be maintained through the pump. With the efficiency known, Δt can be determined from Equation (7). When both the available and required net positive suction head (NPSH) are known, the allowable Δt can be determined. The pump curve and the required NPSH are often not available, however, and the following procedure is suggested as a means of estimating the minimum flow required when the head and flow at the design point (full flow) are known.

The first assumption to be made is that at minimum flow conditions, the entire brake horsepower input is converted into heat. The second is that the brake horsepower near shut-off (minimum flow) is approximately one-half that at the design point:

$$0.5 \times bhp_d \times 2545 = Q_m \times SG \times 500 \times C_p \times \Delta t \tag{8}$$

where

  $bhp_d =$ brake horsepower at design point
  $Q_m =$ minimum flow required, gpm
  $SG =$ specific gravity of the liquid

Also,

  $bhp_d = Q_d H_d SG / 3960 e_d \tag{9}$

where

  $Q_d =$ flow at design point, gpm
  $H_d =$ head at design point, ft
  $e_d =$ efficiency at design point, decimal

These further assumptions are made: efficiency at design point is 0.705; specific heat is 1.0; and the temperature rise is limited to 15°F. Substituting Equation (9) into Equation (8) and solving for Q_m, we obtain:

$$Q_m = 0.000061 Q_d H_d \qquad (10)$$

The temperature rise is added to the temperature of the liquid entering the pump. If the minimum flow is routed immediately back to the pump suction, the same liquid will be recirculated and its temperature will continuously rise. The minimum flow should therefore be routed to a point where it can mix with a large amount of colder liquid, or provisions for cooling the minimum flow should be made.

The simplest method of assuring the minimum required flow is to install a restricting orifice in a recirculation line connected to the pump discharge line. For flow through an orifice,

$$Q_m = 19.64 d_o^2 C_o^2 H^{1/2} \qquad (11)$$

where
  d_o = orifice diameter, in.
  C_o = coefficient of discharge
  H_o = differential head across orifice, ft

For a square-edged orifice, C_o can be taken as 0.61. Further, H_o is approximately the shutoff head of the pump, which can be conservatively taken as $1.1 H_d$. Equation (11) than becomes:

$$d_o = Q_m^{1/2} / 3.55 H_d^{1/4} \qquad (12)$$

The nomograph permits a rapid solution of Equations (10) and (12).

Example. What minimum flow rate and orifice diameter are required if the design flow and head are 450 gpm and 1200 ft, respectively?

Solution. Draw a line from 1200 on the H_d scale through 450 on the Q_d scale and extend it through the d_o and Q_m scales. A minimum flow rate of 33 gpm and an orifice diameter of 0.274 in. are indicated.

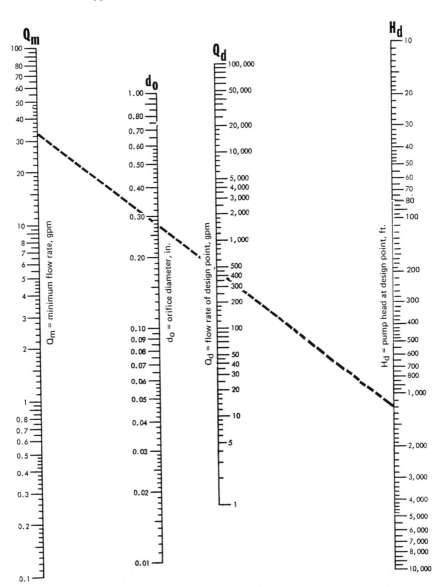

NOMOGRAPH determines minimum flow rate required to limit temperature rise of liquid being pumped, and also the diameter of an orifice to assure minimum flow, when flow rate and head at the design point are known. (Source: *Heating, Piping & Air Conditioning*, Jan. 1966)

8

Pump Operating Principles and Curves

PUMP THEORY REVIEWED

A pump is an energy conversion device, which converts the energy provided by a prime mover such as a motor or turbine, to energy within the liquid being pumped. This energy may be manifested as pressure, velocity head, static elevation, or a combination of these factors.

The rotating element of a centrifugal pump, which is turned by the prime mover, is called the *impeller*. The liquid being pumped surrounds the impeller, and as the impeller rotates, its motion imparts a rotating motion to the liquid (Fig. 8-1).

There are two components to this motion. One component is motion in a radial direction, outward from the center of the impeller, caused by centrifugal force. As the liquid leaves the impeller, it also tends to move in a direction tangential to the outside diameter of the impeller. The actual direction the liquid will take is the resultant of the two components.

The energy added to the liquid by the rotating impeller is related to the velocity with which the liquid moves. Energy, expressed as

136

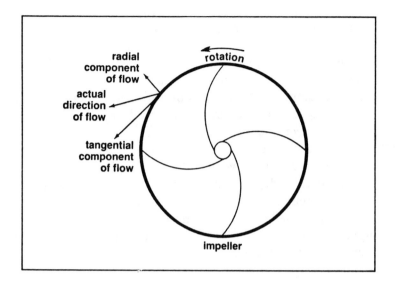

Fig. 8-1

pressure energy, is proportional to the square of the resultant exit velocity:

$$H = K\frac{V^2}{G} \qquad (1)$$

Where
H = energy, ft of liquid
K = a proportionality factor
V = velocity, fps
G = acceleration due to gravity in ft/sec^2

From these facts we can predict two things. First, we can say that anything that increases the tip velocity of the impeller will also increase the energy imparted to the liquid. Second, we can say that changing the vane tip velocity will result in a change in the energy imparted to the liquid which is proportional to the square of the change in tip velocity.

For example, doubling the rotative speed of the impeller would double the tip speed, which in turn would quadruple the energy imparted to the liquid expressed in terms of pressure.

Doubling the impeller diameter would also double the tip speed, which again would quadruple the energy imparted to the liquid.

It should be noted that a change in the tip velocity of the impeller will have the same effect on pump performance whether the change is caused by increased speed or by increased impeller diameter. The pump curves reflect the impeller tip speed.

What happens to the liquid which is being discharged from the tip of the impeller? Taking a volute pump as typical of centrifugals in general, the liquid is discharged from all points around the circumference of the impeller, and moves in a direction which is generally outward from the impeller. At the same time it is moving around with the impeller. It is the function of the pump casing to gather up this liquid and direct it through the discharge nozzle or opening of the pump. The casing is designed so that, at one point, its wall is very close to the outer diameter of the impeller. This point is called the *tongue* of the casing (Fig. 8-2).

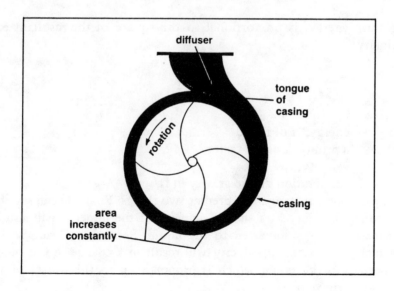

Fig. 8-2

Between the tongue and a point slightly to the left, a certain amount of liquid has been discharged from the impeller. This liquid must rotate with the impeller until it is finally discharged through the outlet nozzle of the pump. Additional liquid is discharged from the impeller at every point around the casing, and this must also travel with the impeller and be discharged through the outlet nozzle. As we continue around the casing, more and more liquid accumulates, which must be carried around between the wall of the casing and the outer edge of the impeller. In order to keep the velocity fairly constant, even though the volume of liquid increases, the area between the tip of the impeller and the casing wall is gradually increased from the casing tongue around to the beginning of the discharge nozzle.

At a point just before the tongue, all the liquid discharged from the impeller has been collected. This liquid must now be led out into the discharge pipe. However, in most cases this liquid possesses a high velocity, which would mean a high friction loss in the discharge piping. Therefore, velocity is usually decreased through the diffuser, by increasing the area for flow. In this way, some of the high velocity energy is changed into pressure energy.

*GENERAL TERMS AND USAGES

Hydraulic Terms and Basic Formulae

Hydraulics is the study of fluids at rest or in motion.

Density, sometimes referred to as specific weight, is the weight per unit volume of a substance. The density of water is 62.3 pounds per cubic foot (at sea level at $60°F$).

Specific Gravity of a substance is the ratio of its density or specific weight to that of some standard substance. For liquids, the standard is water, usually at $60°F$. Specific gravity is a pure, dimensionless number, whereas with density units must be given. In the petroleum industry, API gravity use is: $10°$ API corresponds to a specific gravity of 1.

* Worthington Company, *Power & Fluids,* Vol. III

$$\text{Specific gravity (relative to water at } 60^\circ F)=\frac{141.5}{131.5 + \text{degrees API}} \quad (2)$$

In the chemical industry, degrees Baumé are commonly used. Two scales are used, one for liquids lighter than water, one for those heavier.

For liquids lighter than water,

$$\text{Specific gravity} = \frac{141.5}{130 + \text{degrees Baumé}} \quad (3)$$

For liquids heavier than water,

$$\text{Specific gravity} = \frac{145}{145 - \text{degrees Baumé}} \quad (4)$$

Pressure is the force exerted per unit area of a fluid. The most common unit for designating pressure is pounds per square inch (psi). According to Pascal's principle, pressure applied to the surface of a fluid is transmitted undiminished in all directions.

Atmospheric Pressure (psi) is the force exerted on a unit area by the weight of the atmosphere. The pressure at sea level due to the atmosphere is 14.7 psi.

Gauge Pressure (psig) is a corrected pressure, and is the difference between a measured pressure and the atmosphere; in other words, the pressure scale reads zero at 14.7 pounds per square inch of atmospheric pressure.

Absolute Pressure (psia) is the sum of the gauge pressure and the atmospheric pressure. Psia in a perfect vacuum is 0. Psia at the atmosphere at sea level is 14.7 psi (0 psig).

1 atmosphere (bar) = 14.7 psi = 34 ft of water

$$\text{psi} = \frac{\text{head in feet} \times \text{specific gravity}}{2.31 \text{ (NOTE 1)}} \quad (5)$$

Conversion factor = 34/14.7 = 2.31

Vacuum refers to a pressure below atmospheric. Due to the common use of a column of mercury to measure vacuum, units are most

NOTE 1: This value will vary somewhat, depending on the temperature of the water.

frequently in inches of mercury: 14.7 psi = 30 inches Hg. Note that vacuum can be expressed as inches of mercury below atmospheric, or as inches of mercury above 0.

Vapor Pressure of a liquid at a specified temperature is the pressure at which the liquid is in equilibrium with the atmosphere, or with its vapor in a closed container. For instance, at sea level in atmosphere, the vapor pressure of water at 212 degrees F is 14.7 pounds per square inch, the same as the atmosphere, and therefore non-boiling water is in equilibrium in an open vessel.

At pressures below the vapor pressure at a given temperature, the liquid will start to vaporize due to the reduction in pressure at the surface of the liquid.

Head is a term usually expressed in feet, whereas pressure is usually expressed in pounds per square inch. The formulae for conversions are given above. Actually, the common term "pump head" covers a variety of conditions.

The term "head" by itself is rather misleading. It is commonly taken to mean the difference in elevation between the suction level and the discharge level of the liquid being pumped. Although this is partially correct, it does not include all of the conditions that should be included to give an accurate description.

In general, a head may have three kinds of energy:

1. **Potential Head** is energy of POSITION (measured by the vertical height above some frame of reference).

2. **Static Pressure Head** is energy per pound due to pressure (measured by Bourdon pressure gauges or equivalent means).

3. **Velocity Head** is the head needed to accelerate the liquid. Knowing the velocity of the liquid, the velocity head loss can be calculated by a simple formula: Head = $V^2/2g$ in which g is acceleration due to gravity or 32.16 ft/sec. Although the velocity head loss is a factor in figuring the dynamic heads, the value is usually small and in most cases, negligible. See table:

Velocity Ft/Sec.	4	5	6	7	8	9	10	11	12	13	14	15
Velocity Head-Feet	.25	.39	.56	.76	1.0	1.25	1.55	1.87	2.24	2.62	3.05	3.25

Friction Head is the pressure expressed in lbs/sq in. or feet of liquid needed to overcome the resistance to the flow in the pipe and fittings.

Suction Lift exists when the source of supply is below the center line of the pump.

Suction Head exists when the source of supply is above the center line of the liquid.

Static Suction Lift is the vertical distance from the center line of the pump down to the free level of the liquid source.

Static Suction Head is the vertical distance from the center line of the pump up to the free level of the liquid source.

Net or Dynamic Suction Lift includes static suction lift, friction head loss, and velocity head. Note that velocity head is not counted when suction pressure is measured with a gauge.

Dynamic Suction Head includes status suction head minus friction head minus velocity head.

Dynamic Discharge Head includes static discharge head plus friction head plus velocity head.

Total Dynamic Head includes the dynamic discharge head plus dynamic suction lift or minus dynamic suction head.

Static Discharge Head is the vertical elevation from the center line of the pump to the point of free discharge.

Variations of the discharge head are:

Total Head (H) is the measure of the energy increase per pound imparted to the liquid by the pump and is therefore the algebraic difference between the total discharge head and the total suction head. Total head, as determined on a test when suction lift exists, is the sum of the total discharge head and total suction lift; and when suction head exists, the total head is the total discharge head minus the total suction head.

Total Discharge Head (TDH) is the reading of a pressure gauge at the discharge of the pump, converted to feet of liquid and referred to datum, plus velocity head at the point of gauge attachment.

PUMP PERFORMANCE CURVES

Performance curves for centrifugal pumps are different in kind from curves drawn for positive displacement pumps. This is because the centrifugal pump is a dynamic device, in that the performance of the pump responds to forces of acceleration and velocity. Note that every basic performance curve is based on a particular speed, a specific impeller diameter, impeller width, and fluid viscosity (usually taken as the viscosity of water). While impeller diameter and speed can usually be manipulated within the design of a specific casing, the width of the impeller cannot be changed significantly without selecting a different casing.

The primary curves reflecting pump performance are Head/Capacity, Bhp/Capacity, and Efficiency/Capacity. The capacity, Q, designated in gallons per minute, cubic meters per second, is measured on the abscissa (the X axis) of the plot. The head, H, may be in pounds per square inch, feet of water, or other pressure designations. The head may refer to a complete pump, or in the case of a multistage pump, to one stage only. All stages in a multistage pump usually develop equal heads. Head, Bhp (brake horsepower), and efficiency are measured on the ordinate (the Y axis) of the plot.

Head/capacity, abbreviated as H-Q, is the curve indicating the relationship between the head, or pressure, developed by the pump, and the flow through the pump at any specific head. The curve for a centrifugal pump may slope to the left, to the right, or may be a flattish curve, depending on the specific speed of the impeller. As capacity increases, the total head which the pump is capable of developing is reduced. For pumps, except those having a flattish curve, the highest head occurs at the point where there is no flow through the pump; that is, when the pump is running with the discharge valve closed. The head at the no flow condition is known as the "shut-off head."

The head of a pump is generally expressed in feet and is so plotted on pump curves. This head developed, expressed in feet, is the same irrespective of the fluid pumped. However, the head expressed in pounds per square inch (psi) will be different for fluids of different specific gravity. The heavier the fluid the greater will be the head expressed in psi for a pump. The term "head" is preferred to "pressure," since the former is unaffected by fluid density or temperature.

H-Q curves are plotted for speeds generally corresponding to electric motor speeds at 50 or 60 Hz.

Bhp/capacity is the curve showing the relationship between the power required to drive the pump, and the delivery conditions. As for the head curve, the power curve may be flat, sloping to right or left, or an amorphous shape.

Efficiency/capacity, unlike head or Bhp, cannot be measured directly, but can be calculated by:

$$\% \text{ efficiency} = \frac{\text{head} \times \text{capacity} \times \text{specific gravity}}{3960 \times \text{hp}} \times 100 \qquad (6)$$

Efficiency represents the percentage of useful water horsepower developed by the horsepower required to drive the pump. The difference between the two powers represents losses due to friction, turbulence, bearing and seal friction losses. The water losses translate to heating of the water, which is why a dead-headed pump will develop heating enough to destroy the pump. When dead-headed, all the driver horsepower is converted to heating the water contained in the pump. In working the efficiency formula, remember that 1 hp = 33,000 ft lbs/min., so that the same constants must be used in the formula.

The head and power curves are plotted from tests; the efficiency curve is plotted from the formula.

One other important characteristic which affects pump capability is the Net Positive Suction Head (NPSH). On the pump plot, the $NPSH_R$ is that pressure at the suction necessary to drive the fluid into the eye of the impeller. The NPSH provided by the system, $NPSH_A$, must always be greater than the $NPSH_R$, otherwise the pump will cavitate or lose suction. $NPSH_R$ is usually measured in feet of water, on the right side of the plot. $NPSH_R$ is determined by the manufacturer after extensive testing. Commonly, the stated $NPSH_R$ is that suction head where the pump discharge head is reduced by some two or three percent; that is, the pump is already cavitating to a minor extent. This factor emphasizes the necessity for providing adequate suction head, preferably in excess of that stated on the curve.

By plotting all these characteristics on one coordinate system, the capabilities and limitations of the pump are completely defined, as shown in Fig. 8-3.

Of course, actual figures for the coordinates of head, gpm, bhp and NPSH must be used.

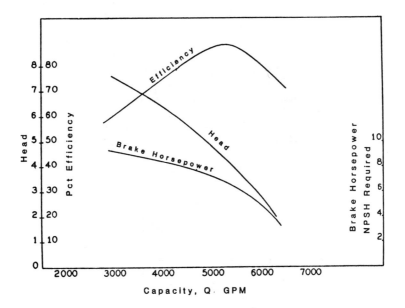

Fig. 8-3

Variation in Pump Curves

All pump curves do not have the sloping curve shown by Fig. 8-3. Fig. 8-4, A, B, and C show nondimensional curves which indicate the general shape of the characteristic curves for the various types of pumps. They show the head, brake horsepower, and efficiency plotted as a percent of their values at the design or best efficiency point of the pump.

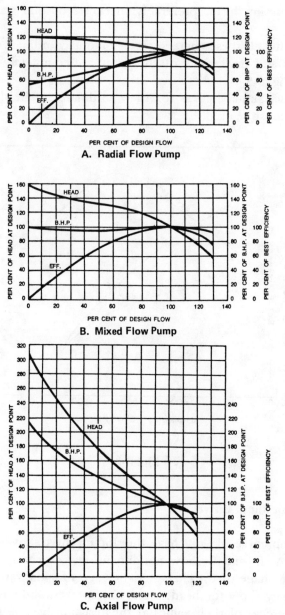

A. Radial Flow Pump

B. Mixed Flow Pump

C. Axial Flow Pump

Fig. 8-4. Non-Dimensional Characteristic Curves for Several Pump Configurations

Fig. 8-4A shows that the head curve for a radial flow pump is relatively flat and that the head decreases gradually as the flow increases. Note that the brake horsepower increases gradually over the flow range with the maximum normally at the point of maximum flow.

Mixed flow centrifugal pumps and axial flow or propeller pumps have considerably different characteristics as shown in Fig. 8-4B and C.

The head curve for a mixed flow pump is steeper than for a radial flow pump. The shut-off head is usually 150 percent to 200 percent of the design head. The brake horsepower remains fairly constant over the flow range. For a typical axial flow pump, the head and brake horsepower both increase drastically near shut-off as shown in Fig. 8-4C.

The distinction between the above three classes is not absolute, and there are many pumps with characteristics falling somewhere between the three. For instance, the Francis vane impeller would have a characteristic between the radial and mixed flow classes. Most turbine pumps are also in this same range depending upon their specific speeds.

Another type of pump, the regenerative turbine, has a typical set of curves shown in Fig. 8-5. These curves are shown for a Roth turbine type pump, operating at 3500 and 1750 rpm. For comparison, curves for a 3500 rpm centrifugal pump are shown by the dotted lines.

In the case of positive displacement pumps, curves are theoretically vertical. Practically, they slope somewhat because of leakage within the pump. The capacity of a vertical pump is determined by its displacement and speed. The pressure is limited only by the construction. The difference between rotary positive displacement pump curves and centrifugal pump curves is illustrated by Fig. 8-6.

Curve Slope

When planning the installation of single or multiple pumps in an installation, the shape of the H-Q curve has considerable influence on the success or failure of the resultant system. It is important, therefore, to analyze the shape of the curve on a particular installation.

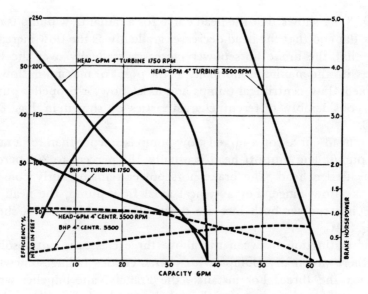

Fig. 8-5. Performance of a Regenerative Type Pump

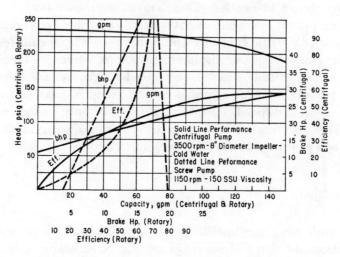

Fig. 8-6. Curves Showing Differences Between
Positive Displacement and Regenerative Type Pumps

Published pump head-capacity curves for centrifugal pumps usually apply to a particular impeller design. Manufacturers can sometimes offer pumps fitted with impellers of different characteristics to meet various applications. Three types of head-capacity curves are characteristic: a "normal" rising curve, a "drooping" curve, and a "steeply rising" curve.

The apparent slope of the curve is influenced by the scales used for the ordinate and abscissa. Therefore, as a method of defining which is which, the difference between the head at BEP (Best Efficiency Point) and at shutoff may be arbitrarily used to designate the type of curve.

In a normal rising curve, or rising head-capacity characteristic, the head rises continuously as the capacity is decreased. See Fig. 8-7A.

The rise from best efficiency point to shutoff may be about 10 to 20 percent. Pumps with curves of this shape are used in parallel operation because of their stable characteristics.

In a drooping curve, the head developed at shutoff is about equal to the head at the BEP (Fig. 8-7B). This characteristic is typical of impellers designed to deliver maximum head per inch of diameter. Efficiency is usually good and the pump may be smaller than normal. When pumps with drooping characteristics are run on throttling systems, they generally operate fine. However, they may cause operating difficulties when the system friction curve is very flat, such as on boiler feed service. These pumps can only be operated in parallel when the operating point is below the shutoff head (developed with discharge closed). An obvious characteristic is that at a given head, shown by a horizontal line on the plot, the pump appears to be unstable because two capacity conditions are possible. This, however, should cause no difficulty as long as the system head curve intersects the pump curve to the right of the second point.

In a steeply rising curve, there is a large increase in head between that developed at design capacity and that developed at shutoff. Fig. 8-7C. This pump will operate in parallel over its entire range. It is best suited for operation where minimum capacity change is desired with change in pressure; for example, in batch pumping or filter systems. The power curve tends to be relatively flat, so the driver is not easily overloaded. The head developed per inch of im-

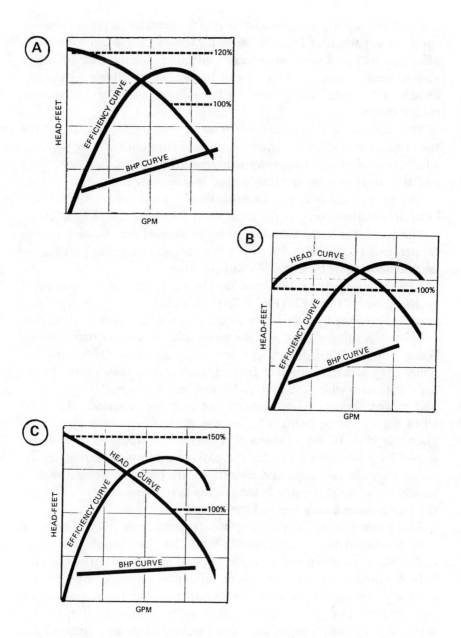

Fig. 8-7. Effect of Various Impeller Designs on Curves

peller diameter is low, so the pump tends to have a lower efficiency and larger size than the normal or flat curve unit.

Fig. 8-8 illustrates that flat curve pumps can be operated in parallel if pumps are removed from the point of intended operation (system curve 2), but not if the pumps are near point of intended operation (system curve 1).

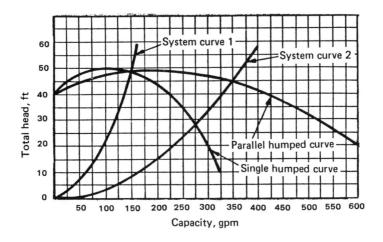

Fig. 8-8. Parallel Operation of Flat Curve Pumps

Curve Selection

When selecting a pump curve to fit the system requirements, it will be convenient to use a composite curve such as that shown in Fig. 8-16. Representing the performance of impellers which vary only in their diameter, and any one of which can be used in a specific casing, selection of the desired operating point becomes merely a matter of inspection. In addition, note that the exact conditions can be satisfied by machining the impellers to a custom size. This is a common procedure, and involves essentially no extra cost. Keep in mind, however, that selection of impellers with the desired steepness, with the operating point near the most efficient part of the curve requires examination of many such composite curves to finally locate the optimum unit. A useful point to consider when looking at composite curves is that these curves are parallel, and therefore can be used to indicate operating parameters at the various speeds that may occur when variable speed drives are used. With a little calculation, the speed necessary to correlate with the change in impeller diameter can be readily determined.

Power curves can show either "overloading" or "non-overloading" characteristics. Speed is a variable which is selected for many reasons: discharge pressure, pump size, quietness of operation, effect of NPSH, and effect on shaft vibration. A 3600 rpm pump does not necessarily wear twice as fast as an 1800 rpm pump; the impeller peripheral velocities for a given discharge head remain constant, as does the rubbing speed on the bearings and seals. The higher speed pump will be smaller, but will generate more noise and require greater NPSH than the slower counterpart.

Shape of the pump curve is important. For parallel operation, a steep curve is best. For constant delivery, a flat curve may be chosen. Shutoff head and runout head and capacity are also defined by the curve. The pump curve is shaped by the design of the impeller.

In an actual centrifugal pump, the impeller is provided with vanes which act to guide the liquid. Furthermore, the impeller has a certain axial width, depending on the capacity it is intended to handle. As an example, Fig. 8-9 shows two impellers of approximately the same diameter, but varying widths. Although their diameters and rotative speeds are identical, the impeller on the left has a flow capability many times that of the impeller on the right.

The design of the casing, being a volute, a double volute, or a diffuser configuration, has considerable effect on the type of loading to which the shaft and bearings are subjected. The arrows in Fig. 8-10 indicate the type of loading that is imposed on the impeller, and thence on the shaft and bearings. In Fig. A, a volute is used; note that the impeller is unevenly loaded, as shown by the arrows of varying length. In a double volute pump, a second volute is added diametrically opposite the first, so that the uneven loadings are balanced and cancelled. In Fig. B, a diffuser is used. It consists of multiple casing guide vanes. Note from the length of the arrows that the loads are quite balanced.

Fig. 8-9

(Source: Worthington Pump, Inc., *Pump World,* Vol. 3, No. 1, 1977)

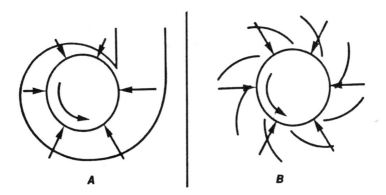

Fig. 8-10

Best Efficiency Point (BEP) is the point on the pump curve where the pump is most efficient. The design operating point should usually be chosen about 10 percent to the left of the BEP. However, there is some thinking that prohibiting operation to the right of the BEP will eliminate many of the possible selections and result in oversizing the pump, and reducing efficiency. A careful study should therefore be made of the projected operation of the entire system.

Impellers and Curves

The numerous pump designs that are available have as a common denominator the impeller. In many cases, manufacturers provide a range of impellers for a single casing style and size. Variations may include several sizes, open or closed, narrow or wide, and various styles of radial vanes, and several impeller eye sizes. In the case of impeller diameter and impeller width, the variations are limited, bearing in mind that efficiency is heavily influenced by clearances between impeller and casing which may exist within the pump. Other variables are allowed more latitude.

Also bear in mind that custom made impellers for minor curve shape changes become prohibitively expensive. Therefore, the impeller choices are really made when the pump is designed, and are not modifications that are available to choose at will. The impeller is therefore implied when the pump characteristics are defined. An understanding of impeller effect on pump performance is therefore useful.

A primary characteristic of impeller design is the closed or open arrangement. The virtues of open vs closed impellers have been discussed in an earlier section. Except for special considerations where solids bearing material is being pumped (or in very small sizes), the closed impeller is preferred. Fig. 8-11 illustrates the cross section of a closed impeller, as well as of typical straight-sided or bevelled open impellers. The latter two, being easiest to machine are often the cheapest configuration. However, two big advantages accrue to the closed profile: reduced side thrust, and usually better efficiency. The efficiency of an open impeller is heavily dependent on the amount of face clearance between the impeller and the casing wall. Although an open impeller can be manufactured with as little as 0.015 inch clearance, it is common for this to enlarge substantially after a short period of service. Many studies, including one made by NASA, have shown this to be the case. Efficiency can drop as much as 10 percent when the clearance is enlarged to 0.050 inch. Closed impeller clearances also widen with extended service; however, the efficiency loss is less. An accelerated wear test of geometrically identical open and closed impellers showed that when the clearances of both impellers opened to 0.050 inch, the efficiency of the open impeller was reduced by 28 percent, but the closed impeller lost only 14 percent. See Fig. 8-12.

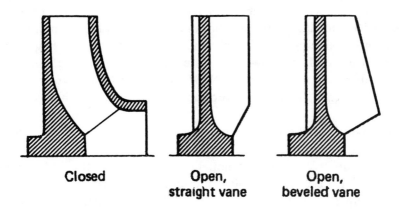

Closed   Open,   Open,
         straight vane   beveled vane

Fig. 8-11

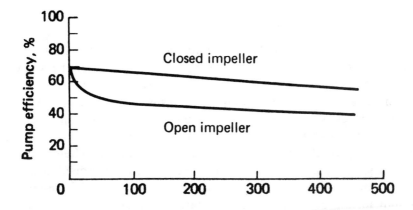

Fig. 8-12

Pump capacity is proportional to impeller width. The wider impeller in a given casing, however, has the disadvantage of a flatter curve than the narrow one. Other changes are brought about by altering the impeller to provide, for example, fewer or more vanes, or more or less "wrap" to the vanes, each of which affects performance. Fig. 8-13. This last difference ties in with another factor in pump design—inlet velocity. If suction conditions are poor, the larger inlet with correspondingly lower velocities will give better performance— i.e.: will work with a lesser NPSH. Condensate and volatile fluid pumps designed for difficult suction conditions usually have oversize inlets.

The width of the impeller in a given casing modifies the efficiency curve. As a matter of fact, in many cases a standard line of pumps will include two groups of impellers: one (called the 100 percent impellers) is designed to utilize the pump casings fully, and the second group (called the 80 percent impellers, for instance) is used to move the best efficiency point to a lower capacity. In this manner, greater coverage is obtained from a line of pumps.

Fig. 8-14 shows the effect of varying the impeller width in a typical pump casing.

Families of Performance Curves

Manufacturers frequently combine all the curves for similar models, but of different sizes, into one graph or plot so that one pump having the correct performance characteristics may be easily selected from the group. One such group of curves is shown by Fig. 8-15. This figure represents a group of ANSI B73.1 horizontal pumps offered by the Ingersoll Rand Company. The numbers on the curves show intake and discharge sizes, and impeller diameter. These numbers identify the pump for further reference to tables or individual curves.

Another curve format is shown by Fig. 8-16, which refers to a pump having a fixed casing size, but able to accommodate various sizes of impellers. The horsepower requirements are shown by sloping lines. Following the curve assigned to a specific impeller size, note that power requirement increases as the capacity increases.

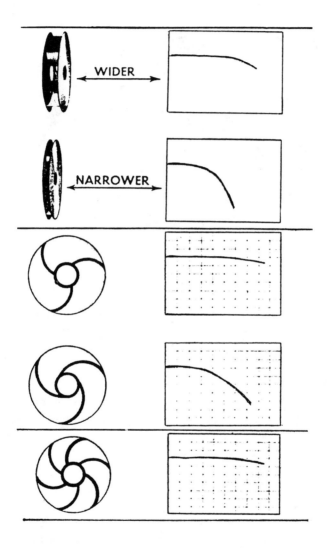

Fig. 8-13. Effect of Impeller Design

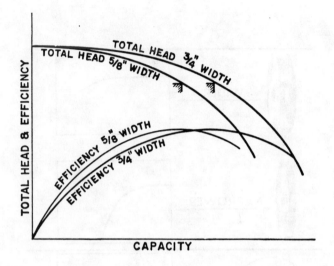

Fig. 8-14. Effect of Varying Impeller Width

Some curves may be plotted with separate power curves for each impeller size. The net result is the same. The efficiencies at any point in the plot are shown by concave curves. The operating point of any pump should be selected somewhat to the left of the center of the efficiency loop. In actual practice, the pump will probably deliver more than the selected gpm. The operating point, moving to the right, will remain at good efficiency, and minimize increase in required motor horsepower.

Pump performance curves are a convenient means of evaluating the important hydraulic characteristics of a given pump. They also serve as accurate guides in selecting a pump which will perform efficiently in a given system.

The pump, being a source of pressure energy, and within its limits of design and operating speed, has the capability of adding energy to the liquid in a system, with increased potential for useful work. The potential energy capability depends on the velocity imparted to the liquid by the impeller. For a specific pump, this velocity can be readily determined by specifying a particular speed, or a particular

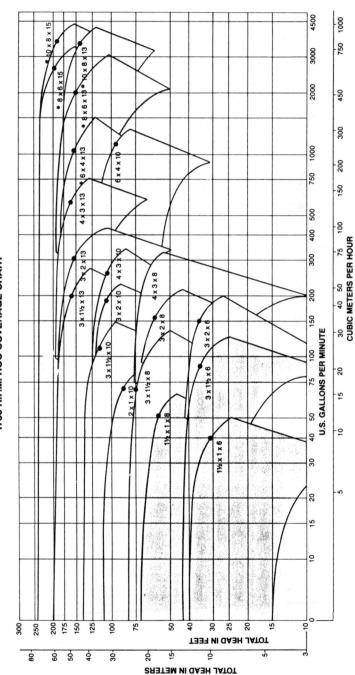

Fig. 8-15

* Open Impeller Only

**50-Cycle Performance Curves and Range Charts available upon request.

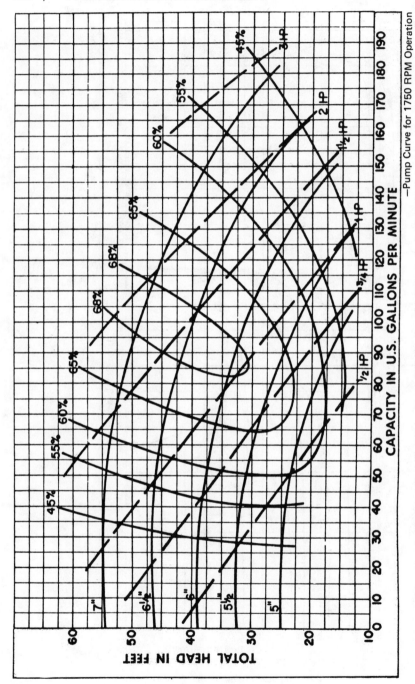

Fig. 8-16. Curves for Various Size Impellers Which May Be Installed in One Size of Casing

impeller diameter. From the curve information available, a centrifugal pump can be used to satisfy a pumping system that has variable head, capacity and efficiency requirements. It should be noted, however, that a driver be selected sufficiently large so that it is not overloaded at any possible operating condition.

9

Similarity Relationships— The Affinity Laws

Any machine which imparts velocity and converts velocity to pressure can be categorized by a set of relationships which apply to any dynamic conditions. These relationships are referred to as the "Affinity Laws." They can be described as similarity processes, which follow the following rules:

1. Capacity varies as the rotating speed—that is, the peripheral velocity of the impeller.

2. Head varies as the square of the rotating speed.

3. BHP varies as the cube of the rotating speed.

The affinity laws apply to centrifugal gas compressors as well as to centrifugal pumps, but are most distinctly useful for estimating pump performance at different rotating speeds, or impeller diameters starting with pumps with known characteristics.

Two basic variations can be analyzed by these relationships:

1. By changing speed and maintaining constant impeller diameter, pump efficiency will remain the same, but head, capacity, and BHP will vary according to the Laws.

2. By changing impeller diameter, but maintaining constant speed the pump efficiency for a diffuser pump will not be affected if the impeller diameter is not changed by more than five percent. Note that the change in efficiency will occur if the impeller size is reduced sufficiently to affect the clearances between the casing and the periphery of the impeller. However, the head, capacity and BHP will vary as follows:

Parameter	Variation 1	Variation 2
	1. With impeller diameter, D, held constant:	2. With speed, N, held constant:
Capacity	A. $\dfrac{Q1}{Q2} = \dfrac{N1}{N2}$	A. $\dfrac{Q1}{Q2} = \dfrac{D1}{D2}$ (1)
Head	B. $\dfrac{H1}{H2} = \left(\dfrac{N1}{N2}\right)^2$	B. $\dfrac{H1}{H2} = \left(\dfrac{D1}{D2}\right)^2$ (2)
Brake Horsepower	C. $\dfrac{BHP_1}{BHP_2} = \left(\dfrac{N_1}{N_2}\right)^3$	C. $\dfrac{BHP_1}{BHP_2} = \left(\dfrac{D_1}{D_2}\right)^3$ (3)

Where Q = Capacity, gpm
  H = Total Head, Feet
  BHP = Brake Horsepower
  N = Pump Speed, rpm

These relationships may be manipulated in any mathematically valid way. The common denominator being a speed change, the relationships may be compiled as follows:

$$Q_2 = (N_2/N_1)Q_1 \qquad (4)$$
$$H_2 = (N_2/N_1)^2 H_1 \qquad (5)$$
$$P_2 = (N_2/N_1)^3 P_1 \qquad (6)$$

A typical calculation using the affinity laws follows:
Assume a "standard" curve showing 100 gpm at 100 psi. The corresponding horsepower is 8 BHP. Consider a "new" curve in which the speed is reduced to 70 percent of standard. The ratio of N_2/N_1 will equal .70. Then:

$$Q_2 \times .70 = 70 \text{ gpm}$$
$$H_2 \times (.70)^2 = 49 \text{ psi}$$
$$P_2 \times (.70)^3 = 2.77 \text{ BHP.}$$

The same procedure may be used when going from 70 percent speed to 100 percent speed. The speed ratio may be taken as 1/.70, or as 1.4285.

Draw the new H-Q curve parallel to the standard curve, going through 70 gpm and 49 psi. The new power curve will be parallel to the standard power curve, going through 2.77 hp at 70 gpm. The new curve can now be used to analyze the system at the new speed or at the new impeller diameter. Use caution when reducing impeller diameter, as the efficiency of the pump may change as the impeller size is reduced. Ten to fifteen percent reduction should yield useable curves when calculating a maximum reduction in impeller size.

Since there are limiting features such as cavitation, shaft size casing strength, etc., it is recommended that the speed be reduced instead of increased. In other words, if it is necessary to operate at 1450 rpm, it is advisable to figure from 1750 rpm rather than from 1160 rpm.

Fig. 9-1 is a typical plot of pump performance of capacity vs head, with the latter being varied by changing impeller diameter. Note that the changed impeller diameters may also be taken as changed speed, using, of course, appropriate scale factors. In common with multiple curve plots, the efficiency curves are roughly parallel; the cusps of the curves may be connected by a nearly straight line.

If the similarity laws are applied to a pump whose impeller diameter is increased, the effects of higher velocity in the system piping must be considered. The laws imply that viscosity will remain constant during passage through the pump. The accuracy decreases as the viscosity increases. Generally, the laws are most accurate when applied to pumps having a low specific speed. When making these calculations, it is always more accurate to go from a higher speed to a lower.

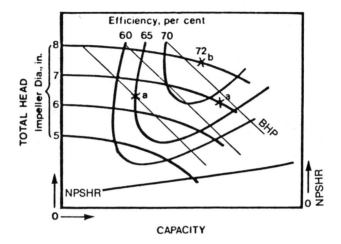

Fig. 9-1. Typical Pump Performance Curve

SYSTEM CURVES

When considering the capability of a pump installation, the primary requirement is to determine the kind of system which the pump is to satisfy. The system curve represents the pressure condition in a piping system, plotted against the gpm flow.

A system curve represents the head required to pump a given quantity of liquid through a piping system. It is essentially a logarithmic curve in which the system resistance increases as the square of the flow. The following relationship applies:

$$\left(\frac{Q_2}{Q_1}\right)^2 = \frac{h_2}{h_1} \qquad (7)$$

Where:
Q_1 = known (design) flow
Q_2 = final flow

h_1 = known (design) head

h_2 = final head

The calculations are not needed to establish a system curve. This is because pipe friction drop varies in a mathematical ratio with the change in water flow rates. Head will change as the square of the change in water flow.

If a slide rule is available, the H-Q relationship can be readily inspected over a range by a simple setting of the slide (Fig. 9-2).

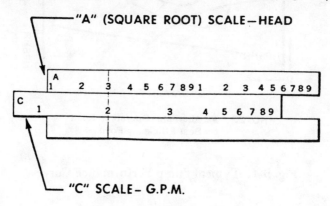

Fig. 9-2

Preliminary calculations provide the known head and gpm values. Set the known head on the A or square root scale. The slide is adjusted so the gpm on the C scale is immediately below the known head. The head may then be read on the A scale for any value of the gpm shown on the C scale. Of course, with the widespread use of pocket calculators, it is very simple to find similar relationships.

The pump performance curve, showing the capabilities of the pump is readily available; however, at this point, it has no real meaning. The pump cannot operate except in conjunction with a system—so it is necessary to coordinate the requirements of the system with the capabilities of the pump. Only then can the selection and application of any specific pump be justified. A system curve must be prepared, showing the conditions which must be satisfied by the pump or pumps to be applied.

The system curve is plotted on the same graph as the pump curves. It represents the head losses, measured on the ordinate, against the capacity, measured on the abscissa of the plot. The system curve represents two factors: (1) the frictional loss through the piping, valves, and equipment; (2) the pressure conditions at the suction and the discharge side of the pump, which are entirely separate from frictional losses. Suction and discharge conditions may be constant, or they may vary with the quantity of fluid being pumped into, or out of the source and the sink. Frictional losses are caused by pipe friction, turbulence, resistance of valves and fittings, and are generally proportional to the velocity squared of the fluid at any point in the system. Although both system and static conditions may be lumped together, it is important to separate specific frictional losses from suction and discharge conditions.

The importance of using a system curve must be emphasized. A system curve intersecting the pump curve shows the head and capacity available from the system. Selection of pumps should not be attempted without this kind of analysis. One additional factor must be considered: system curves almost never start at zero pressure. In a closed system, static pressure may be imposed by the pressure existing in a surge or expansion tank. This would be the starting point for the system curve.

In an open system the static pressure is determined by the elevation of the fluid surface above the datum or zero head on the ordinate scale.

Fig. 9-3 is a simple Head-Capacity plot. The system head curve starts at the static pressure, S1 in a flowing system, the upward trend of the curve represents frictional resistance to flow. The dotted curve represents the same system, showing the effect of a valve closure. Each readjustment of the valve will result in a change to the system curve. The setting of the valve modifies the frictional resistance of the system, and thus provides flow control as required.

Aside from the frictional response of the system to various flows, we can readily identify numerous conditions of suction and discharge which must appear on the plot. For instance: constant head at the inlet, as when pumping from a reservoir; constant head at the discharge, as when pumping into a reservoir; a steadily reducing head, as when pumping out of a tank; a steadily increasing discharge head, as when pumping into a standpipe.

These and other conditions, and any combination thereof, must be carefully analyzed to assure predictable performance of the system.

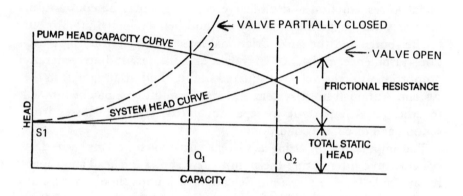

Fig. 9-3. Typical System Head Curve

Fig. 9-4 depicts a simple system, and the appropriate curve for that system. The loss is expressed as a number of feet of head. At zero capacity, of course, there is no friction loss, since there is no flow.

Fig. 9-5 shows the same system and its curve, with the complication that point B is higher than point A. Because of the difference in height, it is necessary to add more energy to the liquid to move it from point A to point B. The amount of energy we must add, expressed in feet, is exactly equal to the difference of elevation between point B and point A. Of course, we must still overcome the friction loss between point A and point B, as we did in the previous system.

The new system can be expressed by the curve shown in Fig. 9-5. The friction curve is exactly the same, because the friction loss between point A and B is the same. But in addition, we must also add a constant amount of head at any capacity—the static difference—just to get the liquid from the elevation at point A to the elevation at point B.

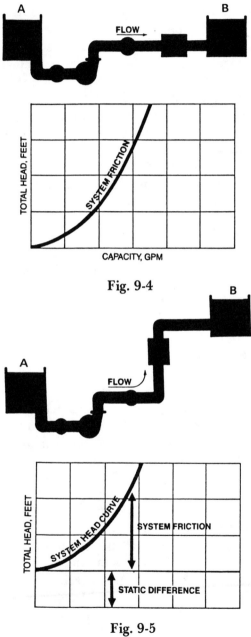

Fig. 9-4

Fig. 9-5

On Fig. 9-5 we were concerned with two different elevations. Instead, the pressure at point A might have been different from the pressure at point B. As an example, if we were taking suction from an open tank at point A and discharging into a closed tank under pressure at point B, it would be necessary to overcome the pressure differential between points A and B. If pressure in the vessel at point B is 10 lb higher than pressure in the vessel at point A, enough energy must be added to overcome this 10 lb, in addition to the friction loss between points A and B.

Many such system variations can exist. In every case, as we are attempting to move the liquid from A to B, there is a friction loss between the two points, and there may also be an elevation or pressure differential, or both.

In a closed system we may take a liquid from point A, move it through a system of piping, and end up at point A again. In such a system there can be no pressure differential between points A and B, since these points are the same. The only loss in the system is the friction loss. The curve for this system, like Fig. 9-4, represents only the friction loss in the system.

PLOTTING THE CURVES

The system elevations must be examined to determine what head differences, if any, may exist between the source and the destination of the pumped fluid. Assume that the flow rate in the system will be 200 gpm, and that the static elevation between points A and B is 20 feet. Friction loss is calculated, using pipe friction tables, and adding in the losses due to equipment or fittings. The friction loss may be 25 feet. The total loss at 200 gpm will then be 45 feet. Doubling the flow to 400 gpm, will quadruple the friction part of the loss from 25 feet to 100 feet.

There are now three points available for plotting the curve: the head at zero flow; the head at 200 gpm; and again at 400 gpm. The curve may be like that shown in Fig. 9-5. Friction loss may not be a fixed value—variable losses must be considered. For instance, a valve might be partially closed, to regulate the flow. The system resistance will increase as shown by the curve of Fig. 9-6.

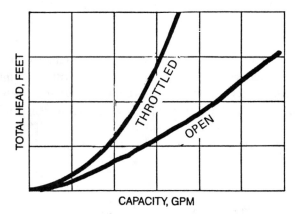

Fig. 9-6. Effect of Valve Closure

Static head may also vary as the pump operates for a period of time. The destination of the fluid may be a tank in which, as it fills up, the level rises. Fig. 9-7 shows the effect of such change on the plot.

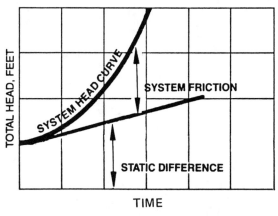

Fig. 9-7

All these variations in a system curve may occur on the same plot. When the system curve is combined with the pump curve, the probable operation of the installation may be studied.

COMBINING PUMP AND
SYSTEM CURVES

With the system behavior defined, it is necessary to select a pump to deliver the necessary capacity, within the limitations imposed by the system parameters. The process requires a flow of 425 gpm, at a head of 132 feet. The nearest curve we can find for pump performance indicates a developed head of 140 feet at 425 gpm. Close enough. Impose the pump curve on the system curve, producing a plot as in Fig. 9-8.

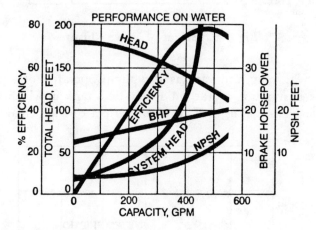

Fig. 9-8

The point of intersection between the pump's performance curve and the system head curve represents the capacity at which the pump will operate. Since the pump does operate at this capacity, brake horsepower required to handle the capacity and NPSH required for proper operation can also be read from the curve. Note that brake horsepower is based on specific gravity of cold water, 1.0. If liquid with higher or lower specific gravity is being handled, determine brake horsepower by multiplying the brake horsepower required on water by the actual specific gravity.

MISCELLANEOUS SYSTEM-HEAD PLOTS

Gravity head: When a system has a negative suction, sometimes called gravity head, the plot will appear as in Fig. 9-9A. If there is considerable positive lift with little friction, the plot may appear as in Fig. 9-9B.

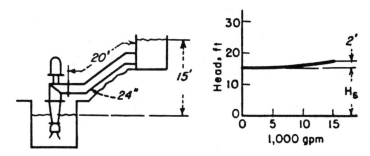

A. Mostly lift; little friction head

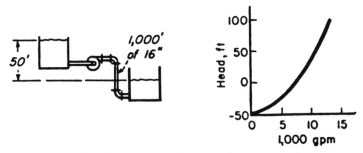

B. Negative lift (gravity head)

Fig. 9-9

Different pipe sizes: When different pipe sizes are found in the same system, the friction loss curve must be plotted independently for each pipe size. For any given flow rate, the total friction loss is the sum of the losses in each of the different pipe sizes. The combined system head curve then represents the static head and friction loss for all portions of the system. Fig. 9-10 applies.

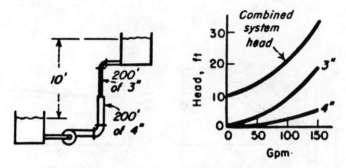

System with two different pipe sizes

Fig. 9-10

Discharge into two different destinations: In many common cases, the pumped liquid is delivered to several locations; for instance, to different tanks, at different levels. The system head curve for each destination must be plotted. Pick off the head loss for each branch, at flow rates which are common for each branch. The combined curve is the sum of the head losses at the specific flow rates. Fig. 9-11 applies.

Flow diverted from the main discharge pipe: The analysis must assume that the flow which is diverted will be a constant quantity. Divide the system into three sections, representing total flow Q_1, diverted flow Q_2, and residual flow Q_3. Plot the friction loss in the normal manner for the pipe up to the diversion point, using Q_1 as the flow. The curve for the residual represented by flow Q_3 should be moved to the right at zero head by an amount equal to flow Q_2, since this is the friction resulting from flows Q_1 and Q_2. The combined system curve, then, will be represented by the sum of the head

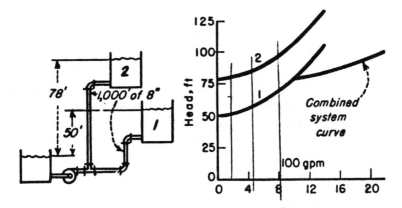

System with two different discharge heads

Fig. 9-11

losses at each given flow rate, plotted as the third curve. The plot should appear as in Fig. 9-12.

Any combination of discharge conditions may be analyzed by dividing the system into sections, each section having a constant loss. The losses of the several sections may then be added, at given flow rates, to result in the combined curve. Some ingenuity may be required, but examination will usually indicate the method of solution.

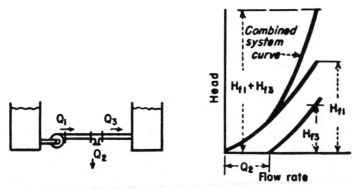

Part of the fluid flow diverted from the main pipe

Fig. 9-12

Pump wear: All pumps wear, and rather quickly develop a performance loss which then stabilizes for a longer period. Manufacturers may be queried as to the change in performance. To "tune" a performance plot, it may be desirable to plot the pump capacities at various flow rates, against the system curve. Note that a flat system curve will maintain better performance from a worn pump than if the system curve was steep. This may suggest that the system should be liberally designed for minimum pressure loss, as shown by Fig. 9-13..

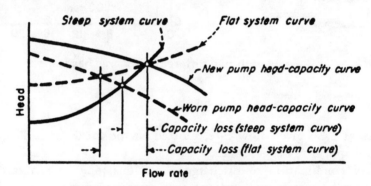

Effect of pump wear on pump capacity

Fig. 9-13

Pumps in series or parallel: The combined curve of pumps in series operation is simply plotted by adding the flow capacities at any given head of each pump, as taken from their curves. Note that there must be considerable similarity in the curves for the combined operation to be satisfactory throughout the range of the pumps.

The combined curve of pumps in series operation is plotted by adding the developed heads of each pump, at any given flow capacity, as taken from their curves. There are a number of limitations to series pump operation. For instance, the second pump should match the capacity of the primary pump, referred to reasonable operating parts of the individual pump curves. Furthermore, the seal and casing design of the second pump must be able to withstand the elevated suction pressure as provided by the primary pump. Again, it should be realized that the primary pump, discharging into the second pump, will experience modifications in its flow and suction conditions which may cause cavitation at the primary inlet. In all, series operated pumps must be carefully matched.

A further discussion of pumps operated in parallel is contained in a later section.

10

Specific Speeds

SPECIFIC SPEED AND
SUCTION SPECIFIC SPEED

Two important theoretical parameters that define the suitability of a pump design for its intended conditions are "Specific Speed" and "Suction Specific Speed." The first term applies largely to "in-pump" operating conditions, and is expressed as:

$$N_s = N(Q)^{0.5}/(H)^{0.75}, \text{ where } N = \text{rpm}, Q = \text{capacity} - \text{gpm, and } H = \text{head-ft.} \quad (1)$$

Specific speed may be further defined as the revolutions per minute at which geometrically similar impellers would run if they were of such size as to discharge one gallon per minute against a one-foot head. The physical meaning of specific speed has no particular value, being a dimensionless number, largely used as a "type" number. It is a constant for all similar pumps and does not change with the speed of the same pump.

The second term applies largely to suction limitations, and is expressed as:

$$S = N(Q)^{0.5}/(NPSH)^{0.75}, \text{ where NPSH is the Net Positive Suction}$$
$$\text{Head, required for cavitation free operation.} \quad (2)$$

178

and

S = suction specific speed—nondimensional
N = pump rotative speed in rpm
Q = pumping capacity in gpm

(Note that with single suction impellers Q is the total flow. With double suction impellers, Q is one-half the total flow.)

The magnitude of the Suction Specific Speed is an "index" of the ability of the pump to operate without cavitation. This parameter is used by pump designers to determine the NPSH requirements of an impeller design.

The specific speed determines the general shape or class of impeller as illustrated by Fig. 10-1. As the specific speed increases, the ratio of the impeller outlet diameter, D_2, to the inlet or eye diameter, D_1, decreases. This ratio becomes 1.0 for a true axial flow impeller.

Radial flow impellers develop head principally through centrifugal force.

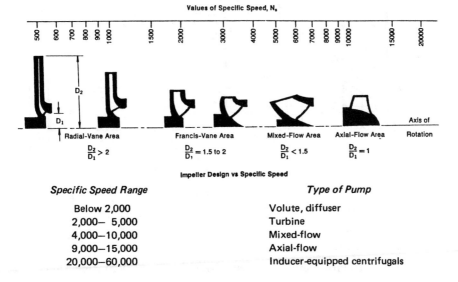

Values of Specific Speed, N_s

Impeller Design vs Specific Speed

Specific Speed Range	Type of Pump
Below 2,000	Volute, diffuser
2,000– 5,000	Turbine
4,000–10,000	Mixed-flow
9,000–15,000	Axial-flow
20,000–60,000	Inducer-equipped centrifugals

Fig. 10-1

Pumps of higher specific speed develop head partly by centrifugal force and partly by axial force. A higher specific speed indicates a pump design with head generation more by axial forces and less by centrifugal forces. An axial flow or propeller pump generates its head exclusively through axial forces. Radial impellers are generally low-flow, high-head design, where axial flow impellers are high-flow, low-head designs.

Fig. 10-2, following, ties together several of the variables that have been discussed. Note that for the example shown on the chart, the required head and capacity are achieved with a corresponding fixed rpm and specific speed.

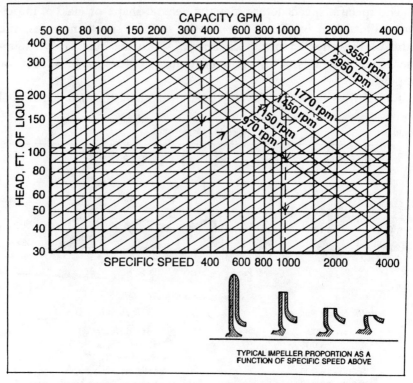

Fig. 10-2. Relationships Between Specific Speed, Rotative Speed, and Impeller Proportions

(Source: Worthington Pump Inc., *Pump World,* Vol. 4, No. 2, 1978)

Increased speeds without proper suction conditions often cause serious problems from vibration, noise, and pitting. The four specific speed curves of Figs. 10-3, -4, -5 and -6 represent the Hydraulic Institute recommended limits of specific speed in respect to capacity, speed, head and suction lift. They should not be construed as theoretical limits. On a particular application it is possible that some pumps will exceed the limits set forth in these curves, where the characteristics of the pump are based on the manufacturers' experience and test data. The curves show recommended maximum specific speeds for normal rated operating conditions, and are based on the premise that at rated condition the pump is operating at or near its point of optimum efficiency.

A new development, not addressed by these charts, is the increasingly common use of suction inducers. These are, in layman's language, "screws" which tend to push fluid into the eye of the impeller. These will be discussed in a subsequent section.

The Hydraulic Institute Specific Speed Limit Charts recommend S values from 7,480 to 10,690, with most of the curves falling below 8,500. These values are relatively conservative and may be raised somewhat, but values of 8,500 to 9,000 should not be exceeded, particularly if the pump is required to operate over a fairly broad range of capacities.

While a first glance examination of the Hydraulic Institute charts would imply that these charts are based on the erroneous concept that total head developed by the pump plays a part in determining the maximum possible rotative speed for a given set of suction conditions, this is not so. The charts are based on constant values or suction specific speeds for different types of pumps. These charts, Figs. 10-3 through 10-6 illustrate the type of information possible.

It is obvious from the suction specific speed equation that a lower pump speed means lower required NPSH. It also becomes obvious that a double suction pump will require lower NPSH than a single suction pump. Fig. 10-7 illustrates "the higher the pump speed, the greater the required NPSH." The chart is based on a suction specific speed of 8,000, which is a sound design value. Actually, where necessary, designs can be made to yield higher values of S, to 10,000 or even higher.

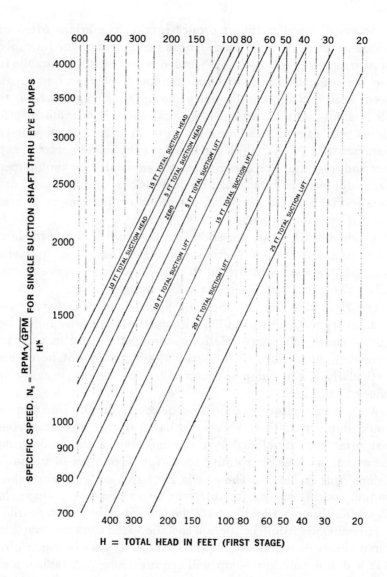

Fig. 10-3. Upper Limits of Specific Speeds for Single Suction Shaft Through Eye Pumps Handling Clear Water at 85° F at Sea Level
(Source: Hydraulic Institute)

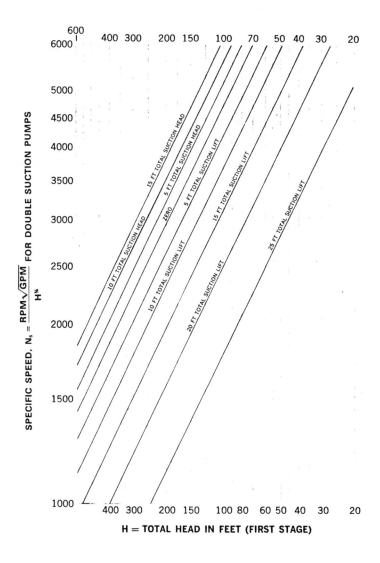

Fig. 10-4. **Upper Limits of Specific Speeds for Double Suction Pumps Handling Clear Water at 85°F at Sea Level**

(Source: Hydraulic Institute)

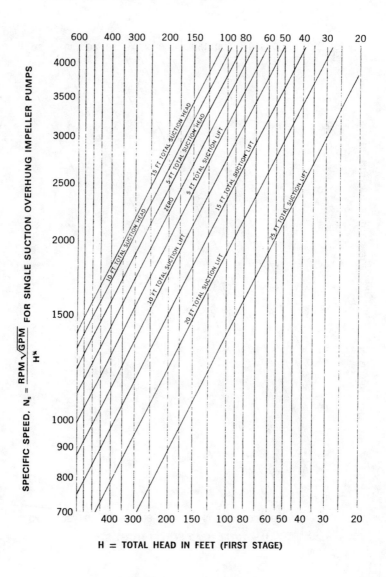

Fig. 10-5. Upper Limits of Specific Speeds for Single Suction Overhung Impeller Pumps Handling Clear Water at 85° F at Sea Level
(Source: Hydraulic Institute)

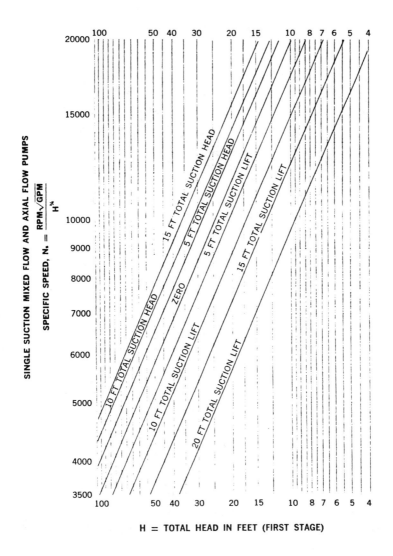

**Fig. 10-6. Upper Limits of Specific Speeds for Single Suction,
Mixed and Axial Flow Pumps Handling Clear Water
at 85°F at Sea Level**

(Source: Hydraulic Institute)

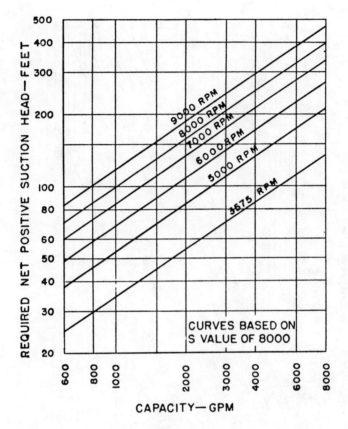

Fig. 10-7

(Source: *Combustion*, April 1963)

If values of S different from those used in preparing the above chart are to be used, correction factors can be applied to the charts to determine new values of required NPSH. These factors are based on the fact that the required NPSH varies inversely with the 4/3 power of S, as shown by Fig. 10-8. This chart is merely a convenient tool to eliminate the need for recalculating NPSH for each new assumed S value. The required NPSH can be established from the formula:

$$H_sS = k \times H_sH, \tag{3}$$

where

H_{sS} = Required NPSH for any desired value of S

H_{sH} = Required NPSH based on Hydraulic Institute Charts in which S= 7,900 for single suction and 6,650 for double suction.

k = Experience factor

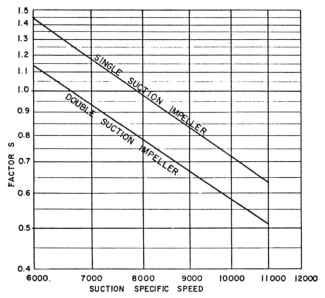

Fig. 10-8. Correction Factor for NPSH
from Hydraulic Institute Charts
(Source: *Combustion*, April 1963)

Specific speed is a correlary parameter to suction specific speed. Specific speed, N_S, is related to suction specific speed, S, by the geometry of the pump. The equation for N_S, repeated here, is:

$$N_S = N(Q)^{0.5}/H^{0.75}$$

In this case, H is the total head in feet of liquid for single stage pumps, or for the first stage of multistage pumps. Note that in this case Q represents the full capacity of the pump, regardless of whether the pump is single or double suction. Fig. 10-9 shows the relationship between N_S and S.

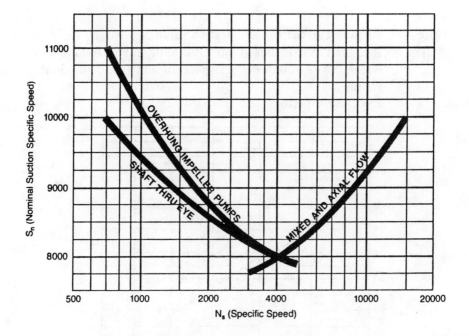

**Fig. 10-9. Relationship Between Specific Speed and
Suction Specific Speed**

(Source: Worthington Pump Inc., *Pump World*, Vol. 3, No. 1)

Fig. 10-10 illustrates the theoretical efficiency of pumps plotted against capacity and specific speed N_S. As will be seen in a subsequent discussion, these curves have an important bearing on the choice of single-stage or multiple-stage designs.

From Fig. 10-2, it can be seen that about the only time efficiency will be at a standoff is when the two pumps of different rotative speeds have specific speeds that are both in the 2000-3000 range. For instance, if the design conditions for the pump are to be 3000 gpm at a head of 100 feet, the specific speed of a pump designed to run at 1770 rpm would be about 3000. For a pump designed to run at 1150 rpm, the specific speed would be just under 2000. The 3000-gpm curve in Fig. 10-2 shows the attainable efficiency for these two

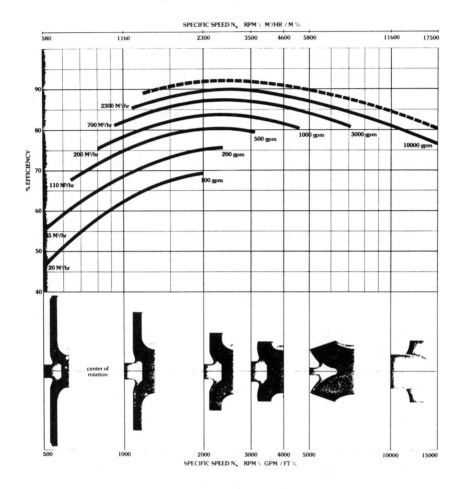

Fig. 10-10

(Source: Worthington Pump Inc., *Pump World,* Vol. 1, No. 1)

specific speeds to be about the same, so that other considerations will govern the choice in this case.

To conclude this discussion of efficiency, consider a situation involving a choice between two commercially available pumps of different rotative speeds. Since no two pumps are normally designed to run at exactly the same conditions, we will use two that are as close as possible. Worthington's 3x2x6 D-1000 pump operates at a speed of 3550 rpm and is designed for 325 gpm at a total head of 150 feet. This condition would correspond to a specific speed of approximately 1600. Worthington also has a 3x2x13 D-1000 pump designed for 350 gpm at a total head of 170 feet, which was designed to run at 1770 rpm. These conditions would correspond to a specific speed of about 740.

Comparing the performance of these two pumps using Fig. 10-2, we find that the 3550-rpm pump has a maximum operating efficiency of 77 percent, while the 1770-rpm pump has a maximum operating efficiency of 67 percent. In other words, for a given set of head and capacity conditions, the 1770 gpm pump will require about 15 percent more power than the 3550 rpm model. Furthermore, the slower motor will be more expensive, and the entire installation will require more space.

An interesting analysis using specific speed parameters is made in *Pump World.** Although efficiency tends to drop off at high specific speed, the greater difficulty is at specific speeds below 1000 (English system notation). In Fig. 10-10 the slope of the efficiency curve below 1000 becomes quite steep, and efficiency falls off rapidly. Therefore, for good efficiency, other factors being equal, it is best to avoid pumps designed for specific speeds below 1000.

Since we are concerned to a large extent with constant speed electric motor drive, 2-pole motor speeds of 3600, or 3000 rpm are normal for many pumps. It is true that variable speed drives are also common on many installations, but the assumption of constant speed simplifies the plot. If a line is plotted for each speed on a field of total head vs flow, representing all the design points for $N_S=$ 1000, a chart as shown by Fig. 10-11 is the result. In this chart, all conditions to the right of the diagonal line have a specific speed greater than 1000; a single-stage centrifugal pump selected for these conditions will run at good efficiency. All applications to the left will have a specific speed less than 1000; efficiency of a single-stage pump for these conditions will be poor.

*Pump World, Vol. 1, No. 1, 1975, Worthington Pump Inc.

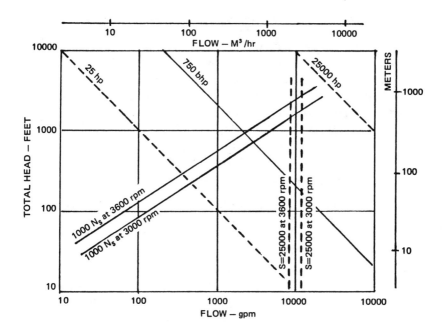

Fig. 10-11
(Source: Worthington Pump, Inc., *Pump World,* Vol. 1, No. 1, 1975)

Immediately to the left of $N_S=1000$ line we have two alternatives that can operate efficiently: multistage pumps with two or more impellers, or higher speed pumps. It next becomes useful to consider the concept of suction specific speed, S, discussed earlier.

While normal centrifugal pump impellers are designed so S is about 10,000 in the English system, with today's technology an S of 25,000 is readily attainable. If we assume that NPSH for many water applications is 30 feet or more, we can locate a line around 10,000 gpm which is the limit for S of 25,000, 30 feet NPSH, and 3600 or 3000 rpm. (Fig. 10-11.) Beyond 10,000 gpm we can still use single-stage pumps with good efficiency, but the driver speed must be reduced to 1750 or 1450 rpm or even lower.

At this point an interesting phenomenon occurs. If we combine the formulae for N_S and S and solve for total head, the result is:

$$\text{Total Head}^{0.75} = S \times \text{NPSH}^{0.75}/N_S \qquad (4)$$

Substituting the limits of S=25,000, NPSH=30 and N_s=1,000 results in a total head of 2,200 ft.

In Fig. 10-11, the diagonal lines for N_s=1,000 intersect the 10,000 gpm line at about 2,200 ft head. The phenomenon is that *any diagonal line for N_s=1,000, at any speed will always intersect the vertical capacity limit of S=25,000 at 2,200 ft head.*

In other words, it is impractical to design a single-stage pump for more than 2,200 ft head, regardless of speed, without either loss in efficiency or increase in NPSH beyond 30 ft.

In Fig. 10-12, additional limits are added, plus a minimum size limit. While this line cannot be defined as precisely as the others, in general there are limitations at speeds over 20,000 rpm, or where impeller inlet is greater than the outlet. There are two other variables deserving of consideration: the inducer and the double suction impeller. The inducer was included in this analysis when the S=25,000 was selected, because this speed cannot be attained without an inducer. If inducers are not used, the S value must be reduced to 10,000 and the maximum flow for 2-pole speeds is reduced to 1300–1800 gpm.

Using a double suction impeller without inducer, the maximum flow for 2-pole speeds would be about 2600–3500 gpm.

Looking at the final results of the plot, as shown in Fig. 10-12, there are several clues provided for pump selection depending on conditions of service. Conditions falling in Area 1 are appropriate for standard single-stage pumps at 2-pole speeds. Area 2 is also covered by single-stage pumps, but at speeds lower than 2-pole. Area 3 provides a choice of either high-speed, single-stage pumps, or multiple-stage pumps at 2-pole speeds. Area 4 is exclusively reserved for multistage pumps. Area 5 is best served with small multistage pumps with as many as 25 stages. Finally, Area 6 is the domain of reciprocating pumps.

Of course, the nomograph is an indication of probabilities, not the last word in pump selection. The analysis does not take into consideration the Kobe Pitot pump, regenerative pumps, or disc/drag pumps.

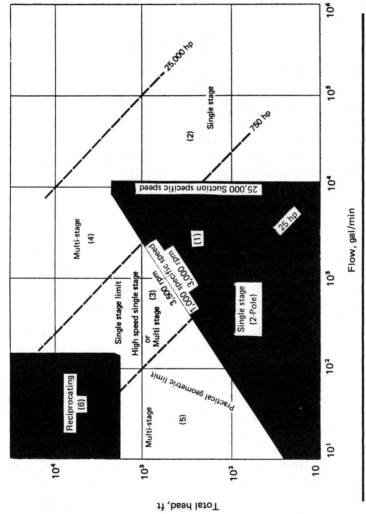

Fog. 10-12

Selection guide is based mainly on specific speed, which number gives an indication of impeller geometry

(Source: *PumpTech*, Vol. 2, No. 2, published by Pump Distribution Association)

In considering specific speed and suction specific speed of pumps, it should not be assumed that very high speeds are not economical or practical. A good example of this is the turbine-driven boiler feed pump in which the turbine wheel and the pump impeller are on the same shaft, separated only by a diaphragm and seal. These units turn at turbine wheel speeds—5000, 6000, 7000 or more rpm. They develop high heads with only one or two stages, and have favorable NPSH characteristics. High-speed, single-stage pumps in the order of 30,000 rpm have been used for years in the aircraft and chemical process industries, driven by high ratio gearboxes from 3600 or 1800 rpm inputs. Pump design principles show that there are three ways of increasing pump head, although combinations of these features may be used.

1. *Increasing impeller diameter,* but at a cost of having a larger, more costly, very inefficient pump. See Fig. 10-13 (a).

2. *Increasing the number of stages.* This is a traditional method of developing high heads, and is very common. The disadvantages are complex and expensive rotating elements with close clearances are subject to abrasion and corrosion. See Fig. 10-13 (b).

3. *Increasing rotational speed.* Fig. 10-13 (c). This involves a relatively small impeller and a speed increasing gearbox. Staging of impellers is eliminated, and the pump size is relatively small. Although the complexity of multistaging is replaced by the complexity of a speed-increasing gearbox, the gears are not exposed to the process liquid, and are extremely reliable. The geometry of a single-stage pump, however, must be matched exactly to the required performance. This type of pump does not forgive off-design operation, or operation below critical speed. A single impeller can readily be tailored to a specific application. Output head can be changed by increasing speed, in which case the efficiency remains nearly constant, provided the impeller geometry remains unchanged. Using a gearbox, the rotational speed is changed in increments using standard gear ratios and the impeller diameter is trimmed to provide the exact head required without affecting efficiency. In this way, pumps can be built with output characteristics anywhere within the envelope shown in Fig. 10-13. Peak efficiency can always be developed at the design point.

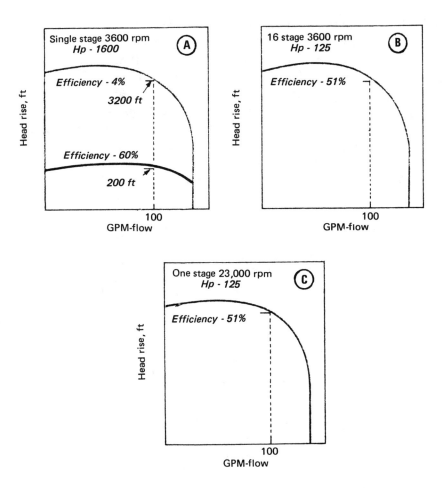

Fig. 10-13. Three Ways of Increasing Pump Output Head
Note significant decrease in efficiency if impeller diameter
is increased at constant speed.

The diffuser bowl is machined to match the impeller configuration and the diffuser cone or outlet is drilled tangentially to the bowl. The liquid is accelerated in the diffuser bowl and then, as it slows down in the diffuser cone, its kinetic energy is converted to the static head required. Fig. 10-14. Because impellers in single-stage pumps are of low specific speed design, very large ratios of impeller diameter to eye diameter can be used. The resulting impeller flow velocity is so low that impeller geometry can be simplified to an open or semi-open design with five or more radial vanes. Pump efficiency is not influenced by impeller axial clearances as great as one-eighth inch. This means that the pump can run dry for limited periods with minimum damage, provided that the seal is suitably flushed. Superspeed pumps used in boiler feed service have proven to be very reliable. Fig. 10-14 shows a cross section of the pump developed by the Sundstrand Corporation.

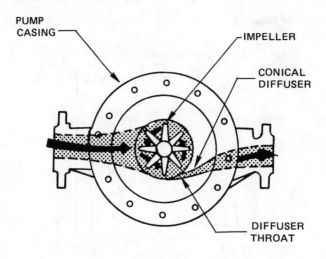

Fig. 10-14. Cross Section of a Single-Emission-Point Pump
(Source: Sundstrand Corporation)

The Sundyne (Fig. 10-14) is a single-emission-point type centrifugal pump, delivering optimum efficiency at high-head, low-flow. For a given speed and diameter, the open, radial-vaned impeller, generates a higher head than a backward-swept impeller. A single divergent conical diffuser efficiently recovers velocity head developed by the impeller. Pump efficiency is not affected by impeller axial clearances as great as one-eighth inch. This design allows the pump to be run dry without damage to the seal if a suitable flush is provided. Lubrication by the pumped liquid is not required. Each pump is sized for the exact flow required, with peak efficiency normally at design point. This is accomplished by sizing the diffuser throat which controls the maximum flow rate. Since it requires only a simple change of drill size, maximum standardization of parts is achieved.

11

Net Positive Suction Head (NPSH)

The centrifugal pump is perhaps the most widely used of any type of equipment in the chemical and process industries, and at the same time, its principles of operation are probably the least understood. Most hydraulic problems occur on the suction side. It is therefore mandatory to understand the factors which must be coordinated to achieve satisfactory operation of the pump system, implying a due regard for the optimum balance of suction system variables. This coordination, or balancing, is an art which can be mastered if due consideration is given to the design, selection and application of the pump.

Commonly, engineers refer to NPSH as the sole governing criterion for the system. Unfortunately, NPSH is not completely established by the conventional formulae. Suction specific speed, cavitation potential, impeller design, clearances in the pump, all have significant effect on the suction requirements. Any discussion of the operation of a centrifugal pump may be conveniently divided into two categories: (1) the discharge or pressure side, and (2) the inlet or suction side. The operation of the discharge side is pretty well understood, as about all that is involved can be covered by the laws of centrifugal force. We know that $NPSH_R$ is the suction requirement established by the manufacturer, and is therefore an integral factor in the selection of the pump. $NPSH_A$, available suction head, is a factor largely defined by the system, and is best calculated when understood. For an application to be completely satisfactory, the $NPSH_A$ must be larger than the $NPSH_R$, the more so the better. However, the operation of the pump on the inlet or so-called suction side is not so well understood.

In fact, one valid criticism of the published theory of these pumps is the fact that it has failed to consider each side separately. It seems to be commonly thought that a foot of head on one side or the other has the same overall effect, but nothing could be further from the truth. The laws of proportionality used to predict the results of change in revolutions per minute on quantity, head, bhp, and efficiency, hold reasonably well for the discharge side of a pump, but not at all for the so-called suction side. Therefore, attempting to compute by the laws of proportionality the performance on the suction side of the pump is an exercise in futility. Tests have clearly indicated that the laws of proportionality do not apply at all to the operation on the suction side. It is clear that there is a definite limit to the head that can be produced on the suction side of the pump. The maximum value is:

$$H_a - H_v \tag{1}$$

when H_a = head available, and H_v = vapor head of a liquid and this value can only be approached. It is also clear that no matter how fast you run the pump this limit still holds. It is equally evident that no more water can be discharged than the pressure of the atmosphere (assuming an open suction source) can force into the eye of the impeller. This would prevent the discharge from increasing indefinitely with increased speed.

Thus, both Q and H are limited by the laws of nature which effectively prevent application of the laws of proportionality. In practice Q can increase with revolutions per minute only so long as the pressure of the atmosphere can supply the required amount of water into the eye of the impeller. Beyond this, some additional positive head must be supplied if flow is to be increased or even maintained. This gives rise to the term "net positive suction head" and represents the actual amount of absolute pressure required at the entrance to the pump to maintain a desired rate of flow.

At this point it might be more exact to simply call this net positive head (NPH) rather than net positive suction head (NPSH), the full meaning of which will be better understood later.

To begin any discussion of NPSH, it is necessary to understand that the term may have two meanings: (1) NPSH available, and (2) NPSH required. Each must be considered separately.

In this connection, it is also necessary to remember that fluid pressure may be expressed in three ways as (1) gauge, (2) absolute, and (3) vacuum. A pressure gauge reads pressures above that of the atmosphere while a so-called vacuum gauge or suction manometer reads pressures below atmosphere. When a gauge indicates a pressure above atmosphere, it is called "gauge pressure." The absolute pressure would be the sum of the gauge pressure and the atmospheric pressure. When pressure exists below that of the atmosphere it may be designated in one of two ways: (1) a vacuum, or (2) an absolute pressure. Vacuum starts with zero at atmospheric pressure and is measured down while absolute pressure starts at absolute zero and is measured upward. All of this is shown in the discussion of calculating NPSH.

The determination of $NPSH_A$ may take different forms, depending on the parameters used for measurement. Referring to Fig. 11-1, the NPSH available in a proposed installation can be calculated by:

$$NPSH_A = H_s = \frac{(P_1 - P_v)2.31}{Sp.\ Gr.} \pm Z_1 - H_{fs} \qquad (1)$$

where

P_1 = Pressure on liquid surface in pounds per square inch absolute (psia). Absolute pressure is equal to gauge reading plus atmospheric pressure. Three common examples are:
1. Open tank: No gauge reading so absolute pressure equals atmospheric pressure or 14.7 psia at sea level.
2. Closed tank under pressure: Add gauge reading in psi to atmospheric pressure to get total absolute pressure.

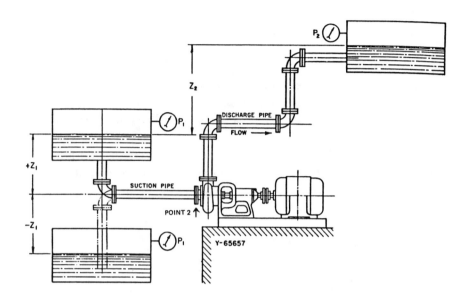

Fig. 11-1. Typical Pump Installation

3. Closed tank under vacuum: Subtract vacuum reading in inches of mercury from atmospheric pressure in inches of mercury (30 inches at sea level) and convert to psia by multiplying by .49.

$$P_1 = (30 - \text{Vacuum}) \times .49$$

P_V = Vapor pressure of liquid in psia at pumping temperature. Available from data tables.

Sp. Gr. = Specific gravity of liquid being handled.

Z_1 = Height of liquid surface above pump suction, measured in feet. If surface is below pump use minus sign.

H_{fs} = Friction loss in feet in suction pipe including entrance loss from tank to pipe, and loss in all valves, elbows and other fittings.

NPSH$_A$ can be calculated in units of feet by omitting the 2.31 conversion factor as follows:

$$H_s = P_1 - P_v \pm Z_1 - H_{fs}$$

where all units are expressed in feet of fluid.

Note that in an existing installation, where suction pressure is measured at point 2 by a gauge, a somewhat different calculation is required.

To determine the NPSH available in an *existing installation*, the following can be employed, in which case it is not necessary to figure out elevations and friction losses because the suction gauge reading accounts for these factors.

$$H_s = P_1 \pm P_s + \frac{V_s^2}{2g} - P_v \tag{2}$$

where:

P_1 = tank pressure or atmospheric pressure for the elevation of the installation expressed in feet of fluid.

P_s = gauge pressure at the suction flange of the pump corrected to the pump centerline and expressed in feet of fluid.

$\frac{V_s^2}{2g}$ = velocity head at the point of measurement of P_s.

P_v = absolute vapor pressure expressed in feet of fluid.

There is a reason why the velocity head term must be considered in formula (2). If the formula for NPSH is based on static elevation differences between the level of supply and the pump centerline datum point, the velocity head cannot and should not be included. The reason for this is that the velocity head results from this static elevation difference, and cannot be counted twice. It is this difference in elevation plus pressure at the suction that causes flow to take place. On the other hand, a gauge located at the pump suction nozzle (point 2) and exactly at the pump centerline would read the static pressure which is a component of P_s, and which lacks the velocity head component to be a measure of the total energy available at the pump suction. Summarizing, the velocity head component *must* be included if NPSH is being determined from a gauge reading at the pump suction, but is already included if the NPSH is established from a difference in elevations. A very elementary consideration regarding fluid supply to a pump is that the pump does not "suck." If there is space available in the pump casing, the liquid is "pushed" into the cavity by the combination of pressures comprising NPSH. If the pressure is not sufficient to fill the casing completely with fluid, the pump will "cavitate." A pump operating in this fashion is said to be operating in the "break." It will cavitate to an increasingly greater amount at all capacities to the right of the break point. It is interesting to note that cavitation can be induced by the introduction of air into the suction, due to air in-leakage, or due to vortices occurring in shallow sumps.

Theoretically, it is possible to lift cold water at 34 feet, corresponding to the sea level atmospheric pressure. However, this lift can never be attained in practice because this 34 feet must be reduced by static suction lift, by friction loss in suction piping, and by shock losses in the impeller eye. The 34 feet of atmospheric pressure must provide for these losses before the fluid vaporizes in the impeller inlet. If the water is warm, so that it has an appreciable vapor pressure, the water will vaporize in the impeller inlet at an earlier point, thus reducing the maximum suction lift. If water or any fluid flashes in the impeller eye, the eye becomes filled with vapor, and prevents any more water being pumped. Sufficient fluid pressure must be available to suppress vaporization, else cavitation occurs.

If available NPSH is less than the minimum required by the pump at the desired capacity, the pump will be unable to meet its head-capacity conditions. A typical group of performance curves for a pump operating under varying suction conditions is shown in Fig. 11-2.

You will note that the reduction in head for any specific suction limitation is not abrupt; in other words, the head-capacity curve does not coincide with the curve with ample excess NPSH up to some capacity and then break off suddenly. Partial cavitation starts at some capacity lower than the complete breakdown, and the head-capacity curve starts to depart slightly, then more and more, from its normal shape. This operation, at point A of the figure, for instance, may result in some reduction in head from the normal head, having

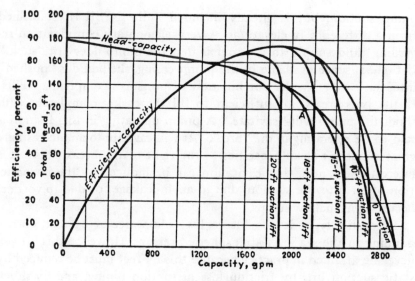

Fig. 11-2

greater NPSH. The determination of critical suction conditions in accordance with the Hydraulic Institute Test Code is referenced to a cavitational coefficient σ (sigma) defined as $\sigma = \dfrac{H_s}{H}$, where H_s = net positive suction head available; H = total pump head per stage in feet.

The net positive suction head requirement for a pump is determined by the manufacturer, following a series of tests. The NPSH is usually set at the point where the pump has lost about 3 percent of its head. However, it should be remembered that a pump operating with the rated NPSH$_R$ is already cavitating.

As a rule of thumb, in noncritical cases, suction lift should be limited to 20 feet, to minimize problems associated with separation (de-aeration or vaporization) and to minimize the need for absolutely airtight suction systems.

The formula for calculating the maximum suction lift a pump can handle is:

$$NPSH_A = H_s{}^2 \, (P_1 - H_V) \, \frac{2.31}{Sp. \; Gr.} - H_{fs} - NPSH_R \qquad (3)$$

where

H_s = suction lift, ft of H_2O

P_1 = atmospheric pressure, psia

H_{fs} = friction loss in ft of H_2O

H_v = vapor pressure of liquid, psia

$NPSH_R$ = NPSH required by pump in ft of H_2O

$NPSH_A$ = NPSH available to pump in feet of H_2O

If more than a 20-feet suction head is indicated by the formula, more sophisticated calculations should be used to determine the precise $NPSH_R$ and $NPSH_A$. The maximum speed at which a pump can be operated for a given capacity depends on the NPSH or the suction lift. The higher the speed the higher the velocity of the inlet vane tip and this causes a greater under-pressure at the inlet requiring more NPSH to operate satisfactorily. Most of the major pump companies belong to the Hydraulic Institute. Its members have prepared specific speed charts based on experience which recommend limiting speeds for centrifugal pumps for various conditions. Refer to Figs. 10-3 through 10-6. On the double suction pump curves where the suction lifting ability of the pump is within the Hydraulic Institute recommended limits for 15-feet lift (the basis for usual guarantees), no limitation curve is shown. On the curves of those pumps which have a lifting ability greater than the Hydraulic Institute recommendations both the Hydraulic Institute limiting curve for 15-feet lift is shown as well as curves showing the actual pump capacities of four pumps with varying suction lift. It is advisable to select the pumps so that under normal operating conditions the Hydraulic Institute limits will not be exceeded. However, it is accepted practice to exceed the Hydraulic Institute limits (but keeping within the maximum pump limits as shown) when the maximum conditions will only prevail for short periods.

NPSH may or may not be a function of impeller diameter. The only sure way to determine the characteristics of a centrifugal pump is by making a series of lift tests from which the NPSH curves can then be calculated. From such tests, we find that some sizes and types of pumps have the same NPSH curve for maximum and minimum impeller diameters; other pumps show different NPSH curves for maximum and minimum impeller diameters. Based on experience in testing a large number of pumps, it is fairly easy to predict the class of NPSH requirements in which a given pump falls.

If the head capacity curves of a pump break at approximately the same gpm, regardless of impeller diameter, the same NPSH curve will apply to the maximum impeller as to the minimum impeller or any impeller diameter in between. (Fig. 11-3A.) Other pump characteristics show a decrease in the maximum capacity with a reduction in impeller diameter. Pumps of this type have different NPSH curves for the maximum and minimum impeller diameter. (Fig. 11-3B.)

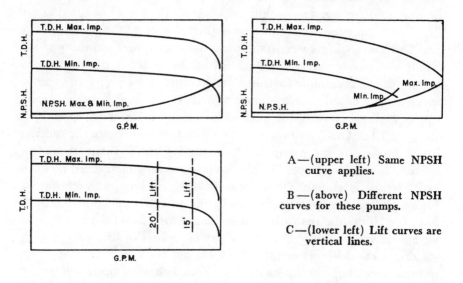

A—(upper left) Same NPSH curve applies.

B—(above) Different NPSH curves for these pumps.

C—(lower left) Lift curves are vertical lines.

Fig. 11-3

Theoretically, the NPSH of a pump is determined by the impeller inlet conditions. The physical dimensions of the impeller eye and vane entrance do not change when trimming an impeller from full diameter to a smaller diameter. Therefore, we could expect that the NPSH at a given capacity would remain the same regardless of the impeller diameter. This is correct within the capacity range from shutoff to about the maximum efficiency point at the minimum impeller diameter. At larger capacities, laboratory tests show that for the same capacity, i.e., for the same impeller eye and entrance velocities, a higher NPSH may be required by the minimum impeller diameter than by the full diameter impeller.

NPSH curves may be different in a variable degree, depending on pump characteristics, for maximum diameter and minimum diameter impeller. Some curves for single-stage double-suction pumps show suction lift lines for maximum capacity at various constant lifts.

For some pumps, these lift curves are straight vertical lines, indicating constant suction lift for a given capacity regardless of impeller diameter. This would mean that the same NPSH applies to any other impeller diameter. (Fig. 11-3 C.) Other curves show a vertical suction lift curve over part of the trim diameters, and at the smaller impeller diameter, the lift curve angles over to the left. That is, the same suction lift at the minimum impeller diameter will occur at a smaller capacity.

12

NPSH Variables

The NPSH calculations must take into account the specific gravity of the fluid, when converting gauge readings to head. The kind of fluid must be considered, since many fluids have low vapor pressure which will cause them to flash readily if subjected to pressures below atmospheric. Finally, the temperature of the fluid, and its resultant effect on vapor pressure, must be considered. A good example is the case of water at 212°F. At this temperature, its vapor pressure is equal to atmospheric, and any pressure reduction caused by suction will simply cause the water to flash. In this case, required NPSH must be obtained by elevating the level of the water sufficiently to provide enough pressure to inhibit flashing.

The relation of NPSH to impeller diameter is not fixed. However, it is fairly easy to predict NPSH requirements in which a given pump falls. If the head capacity curves of a given pump break at approximately the same gpm, regardless of impeller diameter, the same NPSH curve will apply to both maximum and minimum impeller diameters (Fig. 11-3A, page 206). Some pumps show an increase in required NPSH for minimum diameter impellers. This may result from the geometry of the impeller, which has fixed eye diameter, regardless of impeller size. It is an interesting fact, however, that when water is handled at a temperature of 705.4°F, the critical temperature at which no evaporation takes place (as long as pressure is maintained at 3206 psia) the volume of water is the same as that occupied by steam. Therefore, the head-capacity curve of a pump handling 705.4°F water with less than the theoretically required NPSH will coincide with the head-capacity curve using cold water and ample NPSH. Between these two extremes, the relationship between the location of the break and the pumping temperature is a continuous function. Fig. 12-1 illustrates this point. Table 12-1 shows the theoretical lift possible with fluids at various temperatures.

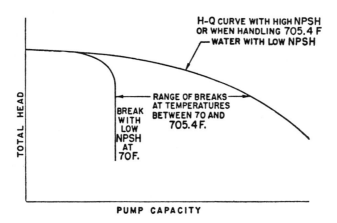

**Fig. 12-1. Effect of Temperature on Maximum Capacity at
Fixed Pump Speed and with Fixed NPSH**
(Source: "Cavitation and NPSH Requirements of Various Liquids," by Victor Salemann,
ASME Paper 58-A-52)

Having established the two limits of performance characteristic at 70° and 705.4°F, we can assume that the relationship between the location of the break and the pumping temperature is a continuous function. Therefore, all breaks at temperatures between these two extreme temperatures must take place between the two limits of capacity indicated.

Table 12-1. Maximum Practical Dynamic Suction Lift and Vapor Pressure

(Source: Marlow Pump Co.)

WATER CHARACTERISTICS

Temp. of	Vapor Pressure PSI Abs	Feet	Specific Gravity	Approx. Maximum Theoretical Suction Lift — Feet	Maximum Practical Dyn. Suction Lift — Feet
40	.1217	0.281	1.0000	33.7	25
50	.1781	0.4115	.9997	33.5	25
60	.2563	0.592	.9990	33.4	25
70	.3631	0.815	.9980	33.1	25
80	.5069	1.17	.9966	32.7	24
90	.6982	1.612	.9950	32.3	24
100	.9492	2.191	.9931	31.4	23
110	1.275	2.942	.9906	31	22
120	1.692	3.91	.9888	30	21
130	2.223	5.145	.9857	288	20
140	2.889	6.675	.9833	27.2	18
150	3.718	8.56	.9803	25.3	16
160	4.741	10.945	.9773	23	14
170	5.992	13.84	.9738	20	11
180	7.510	17.35	.9702	16.5	7
190	9.339	21.55	.9667	12.4	3
200	11.53	26.65	.9632	7.2	2' Positive
210	14.12	32.6	.9592	1.3	8' Positive
220	17.19	39.7	.9552	0	15' Positive

MAXIMUM PRACTICAL DYNAMIC SUCTION LIFTS

GASOLINES • JET FUEL • KEROSENE • SEA LEVEL ATMOSPHERE CONDITIONS

TEMPERATURE	MOTOR GASOLINE				AVIATION GAS	JET FUEL AND KEROSENE
	Winter Gas		Summer Gas			
	Reid Vapor Pressure 14#	12#	10#	8#		
50° F.	15	18	21	23	25	25'
60° F.	11	15	18'	21	23	25'
70° F.	6'	11	15'	18'	20	25'
80° F.	—	5	11	15	18	24'
90° F.	—	—	6	11	14	23'
110° F.	—	—	—	6	10	22'

NOTE: Average temperature of gasoline in underground storage tank is 55° F., not ambient.

APPROX. VAPOR PRESSURE — *PSI Absolute*

FUEL CHARACTERISTICS

TEMPERATURE	MOTOR GASOLINE				AVIATION GAS	JET FUEL AND KEROSENE
	Winter Gas		Summer Gas			
	14 Reid	12 Reid	10 Reid	8 Reid		
40	4.8	4.0	3.4	2.8	2.2	.9
50	5.9	4.9	4.1	3.4	2.8	1.0
60	7.4	6.0	5.0	4.1	3.5	1.0
70	8.9	7.4	6.0	5.0	4.2	1.2
80	10.7	9.0	7.1	5.9	5.1	1.4
90	12.8	10.6	8.6	7.0	6.2	1.7
100	14	12	10	8	7.4	2.0

This would indicate that the practice of assuming that the required NPSH is independent of the pumping temperature is conservative and unrealistic, and that this NPSH is definitely reduced with increased temperatures.

On the other hand, the margin recommended to be added by the Standards of the Hydraulic Institute takes into consideration actual conditions which prevail in a steam power plant—which differ considerably from the "steady state" conditions we assumed in establishing our relation between NPSH requirements and operating temperature. It is well known that "transient conditions" such as sudden load reductions will introduce unfavorable effects on the suction conditions in an open cycle, where the boiler feed pump takes its suction from a deaerating direct contact heater. A very severe reduction in available NPSH follows the sudden load reduction (see *Combustion,* Aug. 1959 and May 1960, "Steam Power Plant Clinic").

Therefore, means must be employed to take care of the time lag which exists between the instantaneous reduction of pressure in the heater which follows a sudden load drop and the ultimate reduction of temperature at the pump suction, after the feedwater already in the suction piping will have been pumped out into the discharge header. The most logical solution is to provide a factor of safety for the installation through the addition of some arbitrary amount to the required NPSH under steady-state conditions.

Returning now to the thought that under steady-state conditions required $NPSH_R$ values are reduced with increased temperatures, it is necessary to introduce a word of caution: it would be highly unwise to rush headlong to the conclusion that the values of recommended $NPSH_R$ for boiler feed pumps should be drastically reduced. We must remember that boiler feed pumps are frequently required to operate over wide ranges of temperature and that a given installation cannot very well take advantage of a permissible reduction in NPSH requirements at the top operating temperature if operation at lower temperatures will wipe out any assistance gained by the effect we have described. The effect is also very useful in affording us additional protection during the transient conditions which prevail when severe load reduction occurs in a steam power plant. Let us preserve this protection and not give in to the temptation to cut corners.

For some time, the Hydraulic Institute has recognized that when the hydrocarbons are handled, minimum required NPSH depends on liquid properties. Worthington has shown that, depending on definition, the NPSH *always* depends on liquid properties.

A certain energy must be contained in the liquid approaching the pump, that will prevent cavitation or its symptoms. In equation form, it is expressed this way:

$$\text{NPSH} = P_{\text{static abs.}} - P_{\text{vapor}} + \frac{V^2}{2g} \text{ (ft)}$$

For a given pump and operating condition, velocity head is constant. If the pump needs a certain NPSH, this value can be obtained by an infinite number of combinations of static pressure and vapor pressure. For example, at sea level, atmospheric pressure on the surface of cold water provides *approximately* 34 ft NPSH. A pump requiring 10 ft NPSH could be placed 24 ft above the water level (neglecting friction losses) and would operate cavitation-free. However, the same pump would have to be placed 10 ft below the surface in an open, boiling tank of water. Here, the vapor pressure at the liquid surface equals the atmospheric pressure; therefore, if the liquid temperature is uniform throughout, the 10 ft total energy above the vapor pressure must be provided for by 10 ft "submergence."

Can a pump running on water at 300° F operate satisfactorily with less available NPSH than the minimum required? The answer is in the definition of "satisfactory" and "minimum required." If "satisfactory" operation is taken to mean negligible hydraulic performance loss, good mechanical performance, lack of noise and cavitation damage, then the pump can operate on less NPSH than the minimum established by the cold water test. Yes, it is cavitating, in the sense that vapor forms at the suction blades and is recompressed into liquid on its way through the impeller. But, due to thermodynamic properties of hot water, only a very small vapor volume is created by a moderate reduction in NPSH below the cold water minimum. Evaporation, however, requires heat. Since this comes from the liquid surrounding the vapor, therefore the whole fluid cools down slightly. Also this temperature drop sufficiently lowers the cavity vapor pressure so that it is no longer below local static pressure—and evaporation stops. Thus, the cavitation process is self-limiting.

In cold water, an infinitesimal drop of available NPSH below minimum required releases a sufficient volume of vapor to affect hydraulic performance. Due to this quick response to a very small pressure differential, evaporation occurs only locally or intermittently. It produces mechanical vibration; and sudden collapse of the cavities filled with only a very light vapor causes noise and mechanical damage. At 300°F, however, a pressure drop of several feet below required minimum is necessary before water develops enough vapor by volume to measurably affect pump performance. This large depression tends to create a more uniform distribution of vapor, allowing the pump to operate smoothly. The relatively high density of the vapor at 300°F and 67 psia slows down bubble collapse as it travels into a high-pressure region, reducing noise and erosion damage.

We conclude that if cavitation is to be absolutely prevented, a pump requires the same minimum NPSH regardless of liquid properties. If, however, the operator will settle for avoiding most cavitation effects (even though cavitation exists in the pumps), then some liquids require less NPSH than others. Since cavitation effects set in with cold water almost as soon as cavitation starts (and since cold water is readily available), other liquids and hot water are usually compared to cold water as a standard.

A condensation of test results, in terms of minimum NPSH for maintaining 97 percent of non-cavitating head, is given in Table 12-2.

LIQUIDS OTHER THAN WATER

The majority of the centrifugal pumps sold today were initially developed and engineered for maximum efficiencies based on accepted water performance. Therefore, when such pump designs are modified to permit handling other products such as slurries, solids, erosive or highly viscous products, they decline sharply in efficiency, gpm, head, and service life. This means that all such pumps must be "derated" in such services compared to their stated water performance.

Table 12-2. Test Results, NPSH Requirements at Design Flow and 3% Drop in Head

(Source: "Cavitation and NPSH Requirements of Various Liquids," by Victor Salemann, ASME Paper 58-A-52)

Fluid	Temperature, F	NPSH, ft, min. accuracy ±0.5 ft	ΔNPSH
Pump N_s = 1600, N = 3585 rpm			
Water	70 250 300	12.3 11.0 8.6	0.0 1.3 3.7
Butane	35 55 90	9.8 8.8 3.5	2.5 3.5 8.8
Butane + 3 per cent propane by weight	35 55 90	9.5 7.8 1.6	2.8 4.5 10.7
Benzene	180 230	12.4 9.7	(−0.1) 2.6
Kerosene degasified	70	12.4	(−0.1)
Gasoline (Reid)	70	13.3	(−1.0)
Freon-11	85 120	10.2 8.4	2.1 3.9
Pump N_s = 1200, N = 3585 rpm			
Water	70 290 325 410[a]	12.0 9.5 6.0 2.0	0 2.5 6.0 10.0
[a] Extrapolated from lower capacity.			

To determine the actual degree of such derating obviously requires exact definition of the material to be pumped.

Pump horsepower is changed with any change in specific gravity. Pump horsepower curves, unless otherwise noted, are plotted for water, which has a specific gravity of 1.00 at normal temperatures. Any increase or decrease in specific gravity will proportionately increase or decrease the horsepower.

Viscous liquids, and liquids heavily loaded with solid material require increased horsepower because of the increase in shear force required and because of increased "drag" of the fluid within the pump. Slurries of many kinds have properties similar to highly viscous fluids. Design for viscous fluids is covered in a subsequent section.

13

Applications of Pump Theory

CAVITATION AND ITS
EFFECT ON NPSH

Just as NPSH is the critical parameter for suction, cavitation is the primary limitation on NPSH. When a pump is cavitating, somewhere within the confines of the pump the pressure will have fallen below the vapor pressure of the liquid at the prevailing temperature. Thus, a small portion of the liquid flowing through the impeller will vaporize, and this vapor will occupy considerably more space within the impeller than the equivalent mass of liquid before evaporation. The term cavitation covers not only the actual formation of vapor bubbles, but also the limitation on capacity, the incidental noise, and the destructive effect on the impeller metal.

All these effects are initiated when liquid pressure at any point on the impeller drops below vapor pressure. This is followed by the re-collapse of these bubbles to liquid when liquid pressure increases to a value just above the vapor pressure. The cause of damage is the shock wave set up by the collapse or "implosion" of these bubbles.

Bubbles initially form at the inlet to an impeller, then are carried along in the liquid stream as it passes through the impeller, until they reach a point where sufficient head has been generated to cause

216

collapse of the bubbles. It is at this point that damage becomes evident. It cannot occur upstream of this point, since in that region there exists a mixture of liquid and vapor, and shock wave energy is dissipated in alternate compression and expansion of the vapor. But in the subsequent liquid phase, the shock wave is propagated through the liquid until it is arrested by the surface of the impeller. If, at the point of arrest, the shock wave possesses sufficient energy, it actually displaces a minute particle of metal from the surface of the metal. Frequent and repeated occurrences of the process produces the pitting which is symptomatic of cavitation. It is evident here that materials more resilient than metal may have a greater degree of resistance to damage. The degree of pitting for a given severity of cavitation is related to several factors: toughness of the metal used in producing the impeller, chemical reaction of the metal with the liquid or with gas held in solution in the liquid and liberated by the reduction in pressure. It should be noted that cavitation and corrosion are complementary causes of damage. The dynamics of damage involve not only the bubble collapse, but also the electrolytic reaction occurring in corrosion. Temperature effects, of course, have a significant effect on the latter.

Serious pitting occurring from cavitation may also affect ships' propellers, so that the problem has widespread implication. It is known that cavitation occurs at "approved" value of NPSH. Indeed, required NPSH is determined by the suction pressure at which the pump under test, has lost from 1 percent to 3 percent of its capacity due to cavitation.

As available NPSH goes down, a point is reached where cavitation appears on the object studied, whether it is a stationary hydrofoil or a moving vane. The NPSH corresponding to the appearance or disappearance of the first vapor bubble, to the first measurable drop in turbomachine performance, or the onset of noise attributed to cavitation, is called "minimum required NPSH." This quantity is a characteristic of the machine and its operating conditions. It is usually considered as independent of the liquid, its pressure, velocity, vapor pressure, and other properties.

A dimensionless cavitation number, K, was developed in an effort to relate the NPSH requirements of a pump to a certain velocity at the impeller inlet. The exact definition of K is, as yet, not finalized. The term appears in the literature in several forms; some researchers use it to express the ratio between NPSH and the relative velocity head of the liquid at the inlet tips. In this case, K is given as:

$$K = \frac{2g\,NPSH}{W_T^2} \text{ where } W_T \text{ is the inlet velocity.}$$

Other analysts replace W_T with U_T, the peripheral velocity of the blade inlet tips.

Generally, K is not as widely accepted as the other terms discussed here. Presently, its use is limited to papers and articles as a theoretical concept.

Considering all the previous discussion about the dangers of cavitation, it seems puzzling that in some cases cavitation is deliberately induced as a means of capacity control. Pumps handling hot water, such as boiler feed and condensate pumps, are much less subject to cavitation than cold water pumps. The potential for interference with pump performance caused by minor cavitation has a definite relationship to the liquid temperature. The definition of cavitation is an implied pressure drop somewhere within the pump, such that the vapor pressure of the liquid at the prevailing temperature is greater than the instantaneous pressure existing at a specific point within the pump. A portion of the liquid will therefore vaporize so that the total space requirement within the pump of the vapor/-liquid mixture is considerably greater than would be required for liquid alone.

In a cold water pump, the volume of a steam bubble is exceedingly greater than the volume of the original volume of water. As an instance, at 50°F, one pound of water occupies 0.016 cubic feet, while steam at the same temperature occupies 2441 cubic feet. The volumetric ratio of these two phases of the same mass of water is 152,000 to 1. As the water is heated, the volume of steam compared to an equivalent mass of water diminishes very rapidly. At 212°F, one pound of water occupies 0.0167 cubic feet, while one pound of steam at the same temperature occupies 26.81 cubic feet. The ratio of volumes at the increased temperature is only 1605, about 100 times less than it was at 50°F. Cavitation at the higher temperature will have less effect than at higher temperatures, due to the markedly lesser quantity of steam comprising the collapsing bubbles.

At 705.4°F, the ratio of steam to water is 1.0, so no cavitation can take place. A slight reduction in the available NPSH can have no appreciable effect on the performance characteristics of the pump. This is illustrated by Fig. 12-1 in the previous chapter. The figure illustrates that the curve for 705.4°F water, with low NPSH, is identical with the curve for cold water and ample NPSH. This explains why, given the appropriate temperature conditions, pump control may sometimes be exercised by throttling the suction of the pump. This, however, should not be considered as a general method of control.

14

Impeller Modification

NPSH VARIATION WITH
IMPELLER CUTDOWN

Current tests on centrifugal pump performance indicate that trim of impeller diameter can have significant effect on NPSH. Worthington Pump Company's tests have indicated the following:

The N_S values are for actual BEP at a given diameter, but in plotting test results, the relation was to flow at BEP with full diameter. This made all comparisons of suction performance of a given pump at two diameters relate to the same flow rate. Two reasons for doing this:

- In regard to the pump inlet, optimum flow rate is constant no matter what the outer diameter. If there is a different flow rate at BEP for a cutdown impeller, the interaction between impeller and volute is the cause.

- Flow rate is perhaps the predominant single factor determining suction capability of an impeller at a given speed. This makes mandatory the use of exactly the same flow rate when comparing performance of a given impeller with various outer diameters.

Often, once a certain flow rate is passed, the required NPSH of a pump with a cutdown impeller increases faster than with the full-size impeller; although many factors contribute to this.

The curves of Figs. 14-1 through 14-5 illustrate the effects of impeller cutdown.

In the past, the assumption has been made that impeller modification does not affect NPSH to any extent. Depending on the pump geometry, this may be in error.

The tests and analyses explain why, at higher flow rates, a pump with cutdown impeller requires a higher NPSH than does one with full-diameter impeller.

The tests also indicate that there must always be a flow rate above which the required NPSH for a cutdown impeller begins to get larger than that for the full-diameter impeller. The flow rate depends on pump design as well as on the percent reduction in diameter.

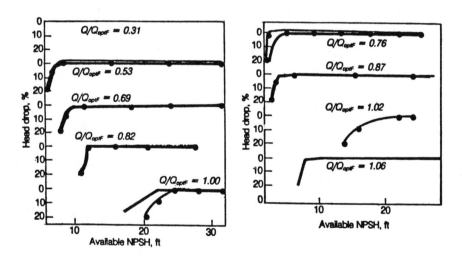

Fig. 14-1 *(left)*. **Test pump 2 began poor NPSH showing at full flow, 3550 rpm, and cut impeller**
(Source: *Power,* October 1985)

Fig. 14-2 *(right)*. **Effect was worse for the pump at 1750 rpm**
(Source: *Power,* October 1985)

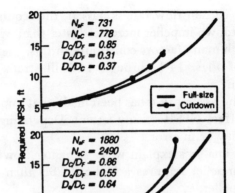

Fig. 14-3. Previous tests on 3550-rpm pumps indicate the effect of impeller trim of about 15%

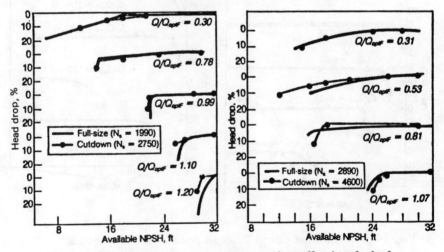

Fig. 14-4 *(left)*. Short-vane test-pump impeller is relatively immune to cutdown at 3550 rpm

Fig. 14-5 *(right)*. Highest N_S test pump, No. 5, also at 3550 rpm, did well with cutdown impeller

(Source: *Power*, October 1985)

In the past, the assumption has been made that impeller modification does not affect NPSH to any extent. Depending on the pump geometry, this may be in error.

The tests and analyses explain why, at higher flow rates, a pump with cutdown impeller requires a higher NPSH than does one with full-diameter impeller.

The tests also indicate that there must always be a flow rate above which the required NPSH for a cutdown impeller begins to get larger than that for the full-diameter impeller. The flow rate depends on pump design as well as on the percent reduction in diameter.

Data from the tests and Fig. 14-2 permit some rough guidelines. As long as flow rate is under Q_{optC}, the NPSH values for the full-size impeller are fairly safe. At higher flows, however, apply those NPSH values with extreme caution.

Leakage through wear rings is another indirect way in which an impeller cutdown may affect NPSH. The leading factor determining required NPSH at a given speed is flow through the impeller. Test instruments do not record this flow rate, however, but rather the sum of that rate and the leakage. Consequently, the higher the leakage, the higher the required NPSH at given net flow rate.

In general, at high flow rates, a cutdown impeller usually has higher required NPSH than does the full-size impeller. In impellers with short blades, however, diameter cutdown can reduce required NPSH.

MODIFYING CENTRIFUGAL PUMPS
FOR NEW CONDITIONS *

When operating conditions change, which in today's HPI plant is more the rule than the exception, pumps must be modified to meet new requirements and/or maintain efficiency. Common changes that can be made relatively fast are:

1) Change driver rpm.
2) Decrease impeller diameter.

*Reprinted from *Petro/Chem Engineer,* January 1968.

3) Modify impeller.
4) Modify casing.

Items 1) and 2) involve fairly simple changes and solve most problems, if not too critical. They shift operating conditions to another curve in the family of curves for pumps of a particular model, usually available from the manufacturer. Items 3) and 4) alter the shape of the entire family of curves.

The pump operating test curve furnished with a new pump is the most useful piece of information on which to base modifications described below. Standard commercially designed pumps are covered here rather than high NPSH or special designs.

The most common method of determining the new operating speed and/or new impeller diameter and required horsepower is through use of affinity law relationships. These are simple to use and results are accurate to about ±15 percent. They assume that efficiency remains constant, whereas it actually decreases about 1 percent per 10 percent decrease in impeller diameter. The new NPSH is also affected, as follows:

$$NPSH_2 = NPSH_1 \ (N_2/N_1)^{4/3}$$

The commonly used affinity laws with their limitations stated above applied to new conditions are as follows:

$$Q_2 = Q_1 \ (N_2/N_1) \qquad (D_2/D_1)$$
$$H_2 = H_1 \ (N_2/N_1)^2 \qquad (D_2/D_1)^2$$
$$BHP_2 = BHP_1 \ (N_2/N_1)^3 \qquad (D_2/D_1)^3$$

Where:

Q = flow rate, gpm
N = speed, rpm
H = total developed head, ft
D = impeller diameter
BHP = brake horsepower

Effects of Changing Driver Speed

Curves for a series of pumps usually cover two speeds, 1750 rpm and 3500 rpm (Fig. 14-6).

When driver speed changes, of course, flow rate, developed head and brake horsepower change also. As pointed out above, using affinity laws to calculate new conditions yields results accurate to about ±15 percent.

Changing motor driver speeds has limited utility since motors come in stock sizes and speeds, but pumpage problems know no stock limitations. Also changing capacity by changing motor speed yields a head change of fixed proportion (square of ratio of speed change), which may not be the desired head.

Precise Impeller Reduction

Reducing impeller diameter to achieve new operating conditions is simple when the original family of curves for the pump series is available. The smallest diameter shown is the minimum size that should be attempted. Trimming below this diameter may pass the point where vane overlap is lost with serious loss of capacity and efficiency.

When using manufacturers' curves, new impeller diameters can be picked without restrictions of traveling lines of constant efficiency. But it is much more complicated when only a test curve is available and the shape of the efficiency envelope is not known.

Calculation of the new impeller diameter from the previously covered affinity law relationships can be modified for greater accuracy by:

1) Reducing efficiency 1 percent for each 10 percent reduction in impeller diameter.

2) Correcting resultant impeller diameter according to Fig. 14-7. This correction chart is most accurate for speeds of 1800 rpm and under.

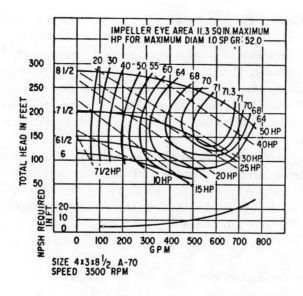

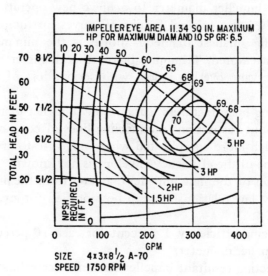

Fig. 14-6. Typical pump family of operating curves at driver speeds of 1750 and 3500 rpm

(Source: *Petro/Chem Engineer,* January 1968)

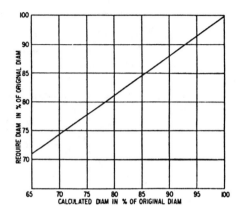

Fig. 14-7. Correction to affinity laws for precise sizing of
impeller diameter

(Source: *Petro/Chem Engineer*, January 1968, courtesy Val Lobanoff, consultant,
Port Washington, N.Y.)

Example

Using the test curve in Fig. 14-8, calculate requirements for new
conditions of 200 gpm at 25 ft TDH (Total Discharge Head). From
desired point draw a line intersecting the pump curve at right angles.
Use this for original conditions.

$Q_1 = 250$ gpm   Efficiency$_1$ = 69.8
$H_1 = 39$ ft
$(D_2/D_1) = (200/250) = (0.80)$
$H_2 = 39 (0.80)^2 = 25$
From Fig. 14-7
(D_2/D_1) Corrected = 0.825   D = 6.2 in.
Efficiency = $69.8 - 1\dfrac{(100 - 80)}{10} = 67.8\%$

New conditions are checked against the family of curves given for
the same pump in Fig. 14-6.

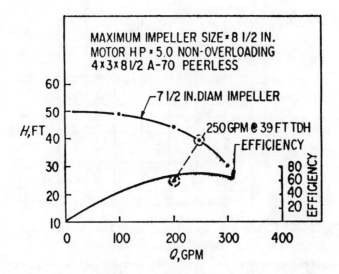

Fig. 14-8. Changing conditions from 250 to 200 gpm at 25 TDH
gives new data for calculations in example problem

(Source: *Petro/Chem Engineer,* January 1968)

These calculated conditions compared with the family curve for
the same pump show:

Calculated Correction		Curve	Affinity Law Not Corrected
Q	250 gpm	250 gpm	250
H	25 ft	25 ft	25
D₂	6.2 in.	6.24 in.	6.0
Eff	67.8	60%	69.8

Utility of the corrections is most obvious in the case of impeller
diameter, which is the most important piece of information.

Effects of Impeller Modifications

New impellers of various types can be purchased if time allows. For example, larger diameters (maximum size is usually given in the original manufacturer's data), thicker impellers, different numbers of vanes, different angles, etc. are available. Of course, when a complete change in impeller design is required, the pump manufacturer should be contacted.

Effect of most modifications can be described qualitatively. The modification of most interest to the engineer manager faced with the necessity of an immediate solution is underfiling of impeller vane tips. The effect of underfiling of impeller tips is to flatten the pump curve and raise the far end of the curve. The shutoff head is unaffected and the efficiency peak is shifted further out on the curve and raised. Fig. 14-9 shows the effect of vane underfiling very clearly. For any given point on the curve, the head remains the same and the capacity is increased as the ratio of the area changes at the impeller tips.

Prior to contacting the manufacturer about changing impeller designs, several generalizations can be made, as follows:

1) Increasing impeller width flattens the curve and increases capacity and width in a manner similar to Fig. 14-9.
2) Decreasing the number of vanes steepens the curve and moves the peak efficiency point down and to the left.
3) Grinding or milling the impeller to reduce thickness retaining the original dimension at the hub and sloping gently to the periphery will affect capacity as shown in Fig. 14-10.

Volute modifications are considered to be difficult to make. Since the casing is the most expensive part of the pump, the average engineer rarely attempts to modify it. Actually, one portion of the casing is very easy to modify—the volute lip or cutwater. This has a similar effect to underfiling the vane tips (Fig. 14-11).

The cutwater often erodes in severe service (Fig. 14-12). When this occurs the jagged keyhole cutout is characteristic. The jagged opening, although increasing area, normally causes turbulence and increased friction and can quite seriously affect throughput and

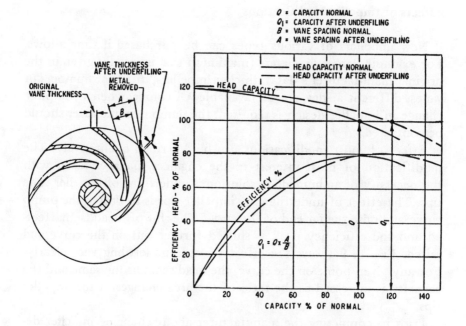

Fig. 14-9. Effect of Pump Impeller Underfiling.
For any given point on the curve, the head remains the same
and the capacity is increased as the ratio of the area changes
at the impeller.
(Source: *Petro/Chem Engineer*, January 1968. Courtesy Val Lobanoff,
Fort Washington, N.Y.)

efficiency. When this happens filing the volute lip back to the furthest
depth of erosion, completely opening the area and leaving no rough
edges, can increase capacity above the original design point (Fig. 14-
11). An often overlooked method for using a high capacity pump for
a low capacity with a high head is using a restrictive orifice on the
pump discharge or external bypass bleeding back to the pump
suction.

The primary difference between the two methods is that bypassing back to pump suction has the effect of decreasing shutoff head

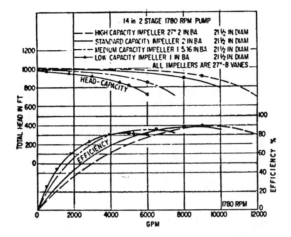

Fig. 14-10. Effect on pump capacity of grinding or milling the impeller to reduce thickness retaining the original dimension at the hub and sloping gently to the periphery.

(Source: *Petro/Chem Engineer,* January 1968. Courtesy Val Lobanoff, Fort Washington, N.Y.)

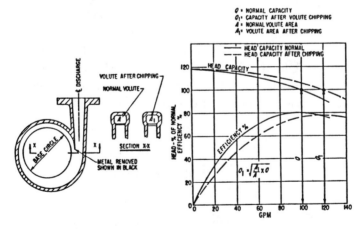

Fig. 14-11. Modifying the volute lip or cutwater has a similar effect to underfiling the vane tips.

(Source: *Petro/Chem Engineer,* January 1968. Courtesy Val Lobanoff, Fort Washington, N.Y.)

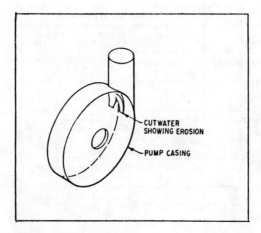

Fig. 14-12. Cutwater showing erosion
(Source: *Petro/Chem Engineer,* January 1968)

without decreasing required horsepower, whereas simply restricting the discharge does not decrease the shutoff head but does lower the required horsepower. Fig. 14-13 shows the effect of a restrictive orifice in the pump discharge. The curve of ΔH and Q is drawn for the orifice. To arrive at the new pump curve, the head loss of the orifice H_O is deducted from the pump head H for the curve required flow rate. The horsepower for the various points should be computed based on the original throughput with the final head desired:

$$\text{Horsepower} = \frac{Q\,H'\,(S)}{3960\,\text{Efficiency}}$$

Where:

$\quad$ Q = throughput, gpm

$\quad$ H' = new pump head with orifice, ft

$\quad$ S = specific gravity of fluid

Efficiency = efficiency on original pump curve corresponding to
$\quad\quad$ Q H

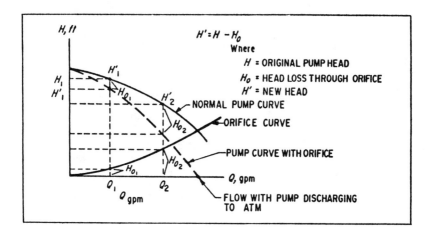

Fig. 14-13. Effect of restrictive orifice in pump discharge
(Source: *Petro/Chem Engineer,* January 1968)

In the case of a bypass orifice, the resistance of the orifice and its piping must be computed in equivalent feet of a given pipe size as shown in Part 1 of this series *(Petro/Chem Engineer,* November 1967, p. 40). The new operating point must be determined in accord with the circuit shown in Fig. 14-14(a), page 234. The equivalent shutoff head exerted on the system will be H_{es} as shown in Fig. 14-14(b) with the corresponding bypass flow Q_{bp}.

A word of caution, if you are operating in less than a crisis condition, contact your pump vendor first before attempting redesign of any pump.

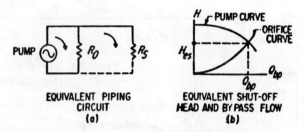

Fig. 14-14. New operating point is determined in accord with circuit on left. Equivalent shutoff head exerted on the system with corresponding bypass flow is shown.

(Source: *Petro/Chem Engineer*, January 1968)

15

Multiple Pump Operation and Curve Slope

Application of pumps in parallel, series, and in combination can provide means of greatly reducing operating horsepower below requirements with conventional pumping methods. These techniques can also reduce installation costs and offer automatic standby pump protection.

In conventional single pump applications, a pump of sufficient capacity to meet design load requirements will typically be installed with another unit of equal capacity to serve as standby. The second pump must be piped, valved, and wired accordingly. With parallel, series, or combination arrangements, the pumps are selected so that their combined capacity will meet design loads with adequate flow maintained when one unit is out for service. This, in many instances, reduces installation costs.

The advantages of using parallel pumps are greater versatility of the pumping system, better matching of the varying load to the pump characteristics, less wear and tear on the pumps, to name a few of the benefits.

By definition, when two or more pumps are installed side by side, connected to a common discharge header, delivering flow simultaneously into the same system, they are said to be in parallel. In parallel operation, the capacities at any given head are added. The shutoff head for the two pumps in parallel is the same as for single operation. Fig. 15-1(a) illustrates the parallel plot for two pumps. The system operation is indicated by adding the system curves (Fig. 15-1(b)). Fig. 15-1(c) is a plot of system and pump curves that indicated a requirement which cannot be satisfied by a single pump. The pumps need not have equal head-capacity characteristics, as long as both can be plotted as shown.

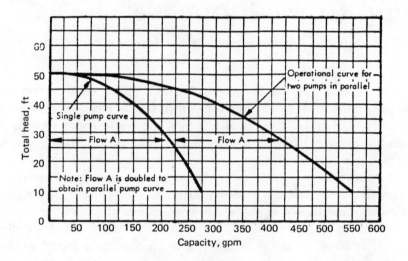

Fig. 15-1(a). In Parallel Pumping, each pump operates at same differential head, supplying one-half the system flow.

When pumps are placed in parallel as shown by Fig. 15-1(a), each pump operates at the same differential head, but each does not necessarily supply one-half of the system flow. The division of flow depends on the relationship of each curve to the system curve. A parallel pump curve can be developed by adding the capacity component of each pump to the other, at a specific head. If the two pumps are duplicates, the flows at constant head can simply be doubled for the plot.

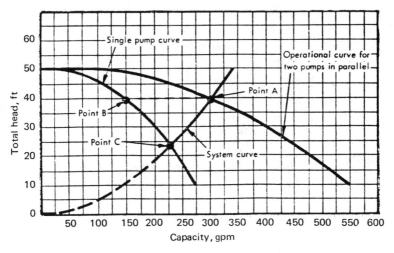

Fig. 15-1(b). Parallel Pump Operation is indicated by adding system curve.

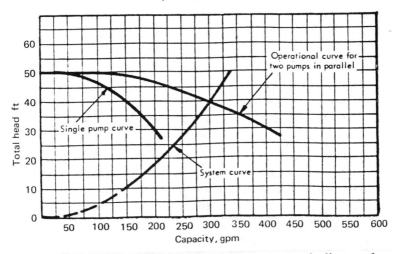

Fig. 15-1(c). Plotting of system and pump curves indicates that both pumps are required; single pump curve does not extend to system curve.

Special care must be taken in selecting pumps for parallel operation. Consideration must be given to single pump operation in the system as well as parallel operation. The first consideration is obvious—placing check valves at the discharge of each pump. Without such valves, during startup and one pump operation, the idle pump will receive a reverse flow which will cause "turbining," and reverse the inertia of the reverse flow of water as well as the reverse rotation. This could lead to motor overload and possible mechanical failure. It may be noted here that single pumps which deliver to an elevated head may also be subject to reverse flow when they are shut down, as the liquid will flow back down when pump shutdown occurs. A check valve is required. This problem is also to be noted in the installation of submersible irrigation type pumps, which are almost always placed below the elevation of the discharge. Since these turbine type pumps cannot be fitted readily with check valves, they are frequently fitted with mechanical reverse rotation stops.

The shape of the pump curves is important, as can be deduced from an inspection of the parallel plot shown by foregoing figures. A special case which should be carefully considered is the inclusion of pumps having a very flat or even worse, a "humped" curve. The latter is shown by Fig. 15-2(a). This curve is illustrative of designs which develop their maximum head at a point to the right of the shutoff point, with a dropping head between the maximum head point and the shutoff point. Inspection of the curve would indicate that the system will not operate properly because for a specific head, there are two capacity conditions. However, the combined pump system plot will usually show that the pump will not operate at two points on the curve, but at the point defined by the intersection of the two curves.

It is frequently thought that pumps with a humped curve should not be operated in parallel. This situation is valid if the humps of the two pumps appear close to the point of intended operation, as shown by system curve 1 in Fig. 15-2(b). However, if the system curve is as shown in the system 2 curve of the figure, operation will be satisfactory.

Reference back to curve of Fig. 15-1(c) indicates a case where satisfactory operation will result only when both pumps operate,

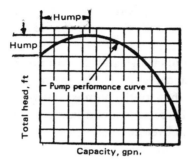

(a) Humped Curve is in itself no drawback; plotting of system
curve will indicate that operation is limited to single point.

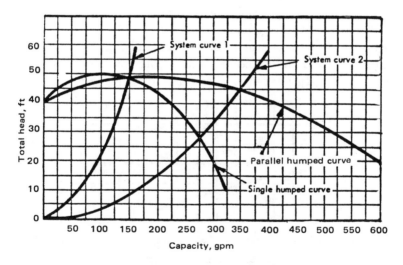

(b) Humped Curve Pumps can be operated in parallel if humps are
removed from point of intended operation (System curve 2), but
not if humps are near point of intended operation (System curve 1)

Fig. 15-2

because the single pump curve does not extend to the system curve. *Operation beyond the end of the published curve must not be permitted.*

When pumps are applied in parallel, with possible single operation beyond the end of the curve, particular instructions must be given to operating personnel that both pumps must always be operated. Pump failure can occur because operating personnel find that single pump operation will deliver adequate flow to satisfy system requirements—not realizing that such operation is beyond the intended pump operating points.

Pumps can be applied in threes, fours, etc., when necessary. Paralell application of line mounted booster pumps can be particularly advantageous. Booster pumps are powered to the end of their curves, and no particular problems are encountered on single pump operation. These pumps, mounted in higher parallel numbers, are often used where economical standby is important, and where "staged" pump flows are advantageous.

Fig. 15-3 shows a complete plot of a parallel pump system, which includes BHP and NPSH requirements. Note that for dual pump operation the NPSH$_R$ per pump is less than for single pump operation.

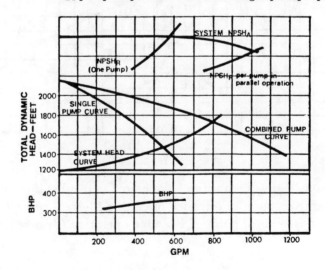

Fig. 15-3

In series pumping, one pump discharges into the suction of the next pump. The pump curves for the combination are plotted by adding the heads at any specific capacity, as shown in Fig. 15-4(a). Application of the system curve to the pump curves is shown by Fig. 15-4(b). As is the case for parallel pumps, any number of pumps may be used.

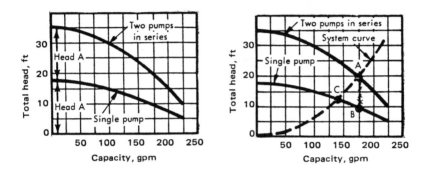

Fig. 15-4(a). *(Left)* In Series Pumping, each pump delivers the same flow rate, contributing half the total head.

Fig. 15-4(b). *(Right)* Pump Operation in series application is shown by plotting system curve on pump curves.

When both pumps are in operation, the total system flow and head will be established at Point A. Each pump will operate at Point B. When only a single pump is employed, the operation is at Point C. Each pump must be powered to Point B, but the power requirement decreases with single pump operation since the operating point shifts to the left on the curve. This is entirely different from parallel operation: a power decrease can be expected with single pump operation in series as opposed to a power increase with single pump operation in parallel.

This particular attribute of series pumping is of great value in varying flow rates and consequently reducing operating costs. On a combined heating-cooling system, for example, the maximum flow requirement is needed only during approximately three summer months. A considerable reduction in flow rate can be established during the winter by simple methods: two pumps can operate in series during the summer, a single pump can operate during the winter.

It is also noteworthy that the NPSH$_R$ is required only to satisfy the needs of the first pump, which will be less than the NPSH$_R$ needed by a single high-pressure pump.

When used for purposes of varying system flow, the pumps should be piped as in Fig. 15-5(a). Either pump can then be valved off to permit service without stopping system circulation.

When pumps are intended only to operate simultaneously, they can be piped as in Fig. 15-5(b). The first pump discharges into the suction of the next. Pumps and water inertia are an aid in starting up since the impeller "turbines" in the correct direction. This is in contrast with the "reverse turbining" of parallel pumping.

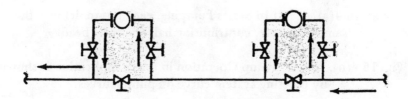

Fig. 15-5(a). Series Pumping arrangement illustrated is to be used for varying system flow rate. By adding a third pump, it constitutes an effective means of providing for future expansion.

Fig. 15-5(b). Series Pumping arrangement illustrated is to be used for simultaneous operation.

In series pumping, each pump delivers the same flow rate but contributes one-half the total pump head. At any constant flow, the head developed by a single pump is doubled, as shown by the curves in Fig. 15-1(a).

It is feasible to combine in one system parallel and series-connected pumps. The plot for such a system requires the parallel curve, and the series curve, as well as individual pump curves and system head curve. The parallel arrangement or the series arrangement may then be selected to economically satisfy a specific loading on the system head curve. Fig. 15-6(a) illustrates such a plot, and Fig. 15-6(b) illustrates a possible series parallel circuit.

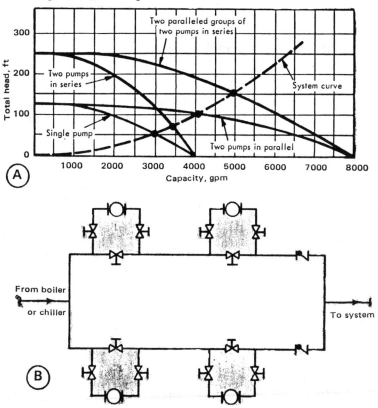

Fig. 15-6

16

Multiple Pump Operation

PROBLEMS WITH MULTIPLE PUMP OPERATION

A study of any system which utilizes multiple pumps will result in avoidance of problems due to poor selection of pump curves. The importance of the degree of slope in the pump curve is critical. The H-Q curves should have two limitation lines drawn on them: one shows the *maximum* flow limit, which most centrifugal pump users are quite aware of; if, for any reason this flow is exceeded, pump enters into destructive cavitation—and perhaps motor overload as well. The other limitation line has to do with *minimum* flow. Few pump users are aware of this limit. (Probably because, in most applications, pump is always operating safely on the right-hand side of this line.) It is the area to the left of this limit line in which noisy low-flow hydraulic turbulence occurs. Destructive forces are at work in this area also, though generally of less severity than those which can literally tear a centrifugal pump to pieces in the low-head, high-flow cavitation area. See Fig. 16-1.

In a system where demand varies from 100 percent to, say, 10 percent, as in the day-night requirements of a public building, the reasonable selection would incorporate a small pump and a large pump for the extremes of load.

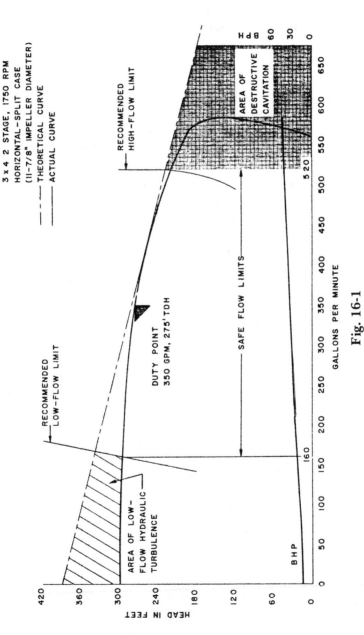

CHART 1

3 x 4 2 STAGE, 1750 RPM
HORIZONTAL-SPLIT CASE
(11-7/8" IMPELLER DIAMETER)
—— — —— THEORETICAL CURVE
——————— ACTUAL CURVE

RECOMMENDED
LOW-FLOW LIMIT

RECOMMENDED
HIGH-FLOW LIMIT

DUTY POINT
350 GPM, 275'TDH

AREA OF LOW-
FLOW HYDRAULIC
TURBULENCE

SAFE FLOW LIMITS

AREA OF
DESTRUCTIVE
CAVITATION

BHP

BHP

HEAD IN FEET

GALLONS PER MINUTE

Fig. 16-1

(Courtesy of SyncroFlo Inc., Norcross, GA)

Exposure to low-flow turbulence is obvious, since the system demand is often less than 10 percent of peak during the night-time hours. So the big pump must cross over into noisy low-flow turbulence. The bigger the pump, the worse is this problem, and the higher is the pump's running cost. A solution would seem to be the use of a small lead pump. Another problem then arises: the danger of high-flow cavitation. This problem can be solved, ordinarily, only by using a pump which is *large* enough to handle the peak flow demand of the system before it crosses the right-hand high-flow limit line.

Theoretically, this danger could still be avoided even if a smaller lead pump were chosen (in an effort to reduce low-flow turbulence and horsepower consumption during light load) by starting a second pump (on pressure drop) before the lead pump gets into trouble. Plotted out on graph paper, this looks perfectly sound. As the system pressure drops to a predetermined point, the next pump appears to be cut in at the right moment. But, occasionally, a control system malfunctions—or perhaps the No. 2 pump develops mechanical trouble which prevents it from running. Or maybe the overload on its driving motor has tripped out. Or a branch fuse has blown. Any *one* of these leaves the No. 1 pump exposed to destructive cavitation. Refer to Fig. 16-2.

At best, a close-set overload trip-out on the No. 1 pump saves the pump itself from distruction, but this, then, leaves the water system with two pumps out of service.

There is another way in which a half-size pump is forced into destructive cavitation. Most designers anticipate a minimum city supply pressure when specifying booster pump duty. But most of the time, normal city supply pressure is above this. It can often be 20 pounds or more above the design minimum. This means that the pump, in most split-pump systems, is started from a pressure signal which can be set, logically, only from a measured value above the anticipated minimum suction pressure. As long as the flow requirement stays within the safe operating range of the lead pump, any rise in suction pressure does no harm at all. But now the flow increases beyond the safe reach of the No. 1 pump and this extra 20 pounds of suction pressure has been lost. But the head curve on this type of pump is too flat, such a pump will usually increase its capac-

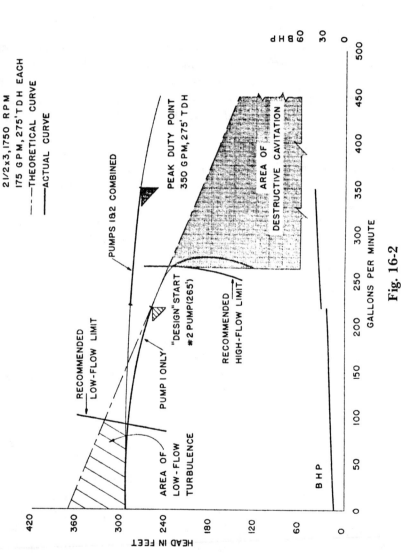

CHART 2

2-50% CAPACITY SPLIT-CASE PUMPS
2 1/2 x 3, 1750 R P M
175 G P M, 275' T D H EACH
– – –THEORETICAL CURVE
——ACTUAL CURVE

Fig. 16-2

(Courtesy of SyncroFlo Inc., Norcross, GA)

ity enormously before dropping its head pressure an extra 20 pounds. The pressure signal waits, satisfied with the fact that the system pressure has not yet fallen far enough to prompt it to cut in the next pump—unaware that the lead pump has overextended itself and has entered into the destructive cavitation area.

If the pump curve has a steep enough head curve, minimum low-flow turbulence occurs, and it would not run out and cavitate at low head. So, if any such pump is to be used in a constant-speed system, the choice of a large pump, at or very close to 100 percent of design capacity, appears to be dictated to the system designer, if the primary danger of destructive low-head, high-flow cavitation is to be successfully avoided. Lower first-cost and minimum system-pressure-droop and elimination of sequencing controls are also in favor of this choice, at the same time that maximum low-flow turbulence and higher running cost must be tolerated.

The foregoing discussion illustrates the importance of preparing system curves and selecting the pumps with correct H-Q curves.

MULTI-STAGE PUMPS

Many centrifugal pump applications can be handled by single-stage pumps, and frequently a single-stage pump is selected without regard to the possible benefits offered by a multi-stage unit. Aside from that, however, a multi-stage pump should be chosen when excessive impeller speed or diameter is required to achieve the necessary head or flow.

It is obvious that a multi-stage pump will be more expensive than a single-stage when only the first-cost is considered. Long-term savings frequently result from increased efficiency, and reduced wear of the moving parts. What criteria, then, should be used to determine the optimum staging for a pump to satisfy given requirements? Possibly the best criterion is the concept of Specific Speed, which has been discussed at length in Chapter 10. A good analysis is given in *Power & Fluids,* the Worthington company's publication as follows:

The centrifugal pump is a hydrodynamic machine with an impeller designed for one set of conditions of head and total head at any given speed. Impeller shape runs the gamut from very narrow large impellers for low flows, through much wider impellers for larger flows, to the specialized propeller for highest flow, low-head conditions. Unfortunately, not all designs can have equally good efficiency. In general, medium-flow pumps are the most efficient; extremes of either low- or high-flow will drop off the efficiency, as illustrated by Fig. 10-10. The best attainable efficiency is a function of impeller geometry, implied by the dimensionless factor called Specific Speed:

$$\text{Specific Speed } (N_S) = N Q^{0.5}/H^{0.75}$$

where

N = pump speed in rpm

Q = flow capacity in gpm

H = developed head per stage in feet of liquid

(Note that in the metric system capacity is given in meters^3/hour and in meters of head.)

While efficiency tends to drop off at high specific speed, the greater difficulty is at specific speeds below 1000.

In Fig. 10-10 the slope of the efficiency curve below N_S=1000 becomes quite steep, and efficiency falls off rapidly. Herein lies the advantage of a multi-stage pump—the multiple impellers permit the use of an overall specific speed above 1000, whereas a single-stage pump for the same head would require a much lower specific speed, with consequent reduction in efficiency. Referring to the curve of Fig. 10-10, it is evident that maximum pump efficiency is achieved at a specific speed of about 2000. Therefore, it is desirable to use N_S=2000 as a target for pump selection and design. Another benefit with the use of multiple impellers is a reduced NPSH$_R$. Pump selection is usually based on operation at or near the Point of Best Efficiency (BEP). This point is determined by an examination of the pump curves. In the formula for Specific Speed, the head is developed per stage. For multi-stage pumps, the total head required is divided by the number of stages. Analysis of pump performance with any number of stages may be tabulated as follows:

Assume a pump head of 3600 feet, 1800 rpm, and a capacity of 600 gpm. Calculate the Specific Speed from the equation, and read the theoretical efficiency from the curves of Fig. 10-10. These values are not necessarily a practical design, but are illustrative.

Number of Stages	Head per Stage, ft	Specific Speed	Efficiency, %
1	1800	322.3	below the curve
2	900	536.7	below the curve
3	600	728.8	70
4	450	908.9	75
5	360	1067.3	76
10	180	1795.6	80
12	150	2055.1	81

Note that for this high-pressure service, anything below three stages should not even be considered. At the upper end of the number of stages, 10 or 12 stages would appear to be optimum. High-pressure pumps are built with 20 or more stages. It seems that numerous stages would lose efficiency because of friction and turbulence, but actually the passages are carefully designed, resulting in smooth flow, with very small friction loss. The method of calculation shown here applies to any multi-stage pump.

The power requirement of the illustrated pump will be proportional to the efficiencies tabulated for the various number of stages. The power will vary from about 564 hp for the three-stage pump, to about 391 hp for the 12-stage pump.

There are some sizing considerations that are loosely dependent on the number of stages. Fig. 16-3 charts impeller diameter vs head, and impeller diameter vs approximate width for a hypothetical pump. These curves may vary with individual designs, but they illustrate general relationships between impeller sizing and impeller developed head. A general relationship exists between impeller diameter and hydraulic width. Diameter/width ratio should be limited to less than 35. A D/W ratio between 35 and 12 represents reasonable proportioning for the impeller. The D/W ratios over 35 may present manufacturing problems, especially with one-, two- or three-stage pumps. Likewise, the volute passages for the impellers at such ratios would be difficult to cast.

IMPELLER WIDTH VS DIAMETER

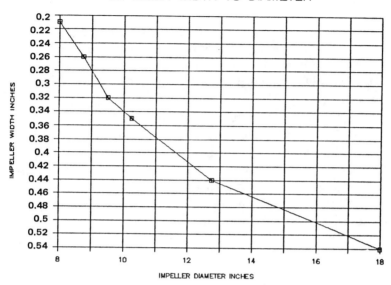

HEAD VS. IMPELLER DIAMETER

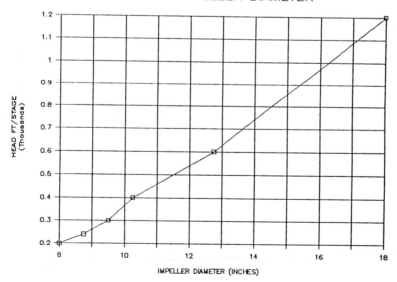

Fig. 16-3

There are other significant advantages attributable to the multi-stage arrangement:

The staggering of volutes on successive impellers at 180° from each other produces equal and opposite loads to balance radial hydraulic forces. The result is a minimum of radial forces on the shaft, allowing the use of a smaller shaft, lower deflection at the seals, and smaller bearings.

High axial thrust characteristic of large impellers is difficult to counteract. The axial forces increase as the square of the impeller diameter. The use of smaller impellers in multi-stage pumps thus greatly reduces axial thrust. The multiplication of thrust forces due to multiple impellers is minimized by using an even number of impellers with thrust forces in opposition to each other, or by having all the impellers facing in the same direction and counteracting the impeller thrust by the use of a balancing drum at one end of the rotor.

Multi-stage pumps may use double-suction impellers which minimize the axial thrust and reduce the $NPSH_R$.

Most significantly, using multi-stage construction, only the first stage is required to meet the necessary suction limitations. The first impeller, designed for only a fraction of the total head, will have reduced suction requirements.

Finally, a small diameter casing is easier to cast without flaws than is the case of the larger casing.

Further discussion of multi-stage pumps will indicate that they are more versatile, and provide greater choice in selection, than is the case with single-stage pumps.

17

Effect of Pump Speed on Selection, Design and Use

A very common conception among pump users is that the slower the pump rotative speed, the better the pump and its operation. The head that an impeller can develop is based on its diameter and its rpm. For a given speed, the greater the diameter, the greater the head and capacity capability of the pump. The speed and impeller diameter are complementary; for a given head, halve the speed and double the diameter; or, double the diameter and halve the speed. It becomes obvious, that for a given set of service conditions, the higher speed pump will have a smaller impeller diameter. The relationship is basically linear. This is a simple relationship, but has a number of concomitant factors that are not always considered:

• The lower speed pump will be larger, and therefore more expensive to buy and install, even though the piping size stays constant. The cost difference may be particularly noticeable if the pump is constructed of expensive materials.

• The efficiency of the higher speed pump is greater, so that a smaller motor may be used. Even if the horsepower is the same, a smaller NEMA frame size will be satisfactory for the higher speed pump.

- The pump shaft and bearings on the lower speed pump will be larger, as will the seals.

- The radial reaction (pressure distribution around the circumference of the impeller) is greater for the slower pump, causing added loads on the bearings. This is a very significant problem, involving not only the cost of larger components, but requiring that the shaft diameter be optimized to reduce deflection at the bearings and at the seals.

- General consideration of the slower speed pump statistically implies that it would have a longer bearing life. This is a general fallacy, as bearings do not fail because of age. They normally do not operate under such conditions that statistical failure is obtained. Rather, over lubrication, under lubrication, contamination or heating cause bearings to fail long before their statistical life has been attained. In such circumstances, the consideration of rotative speed is not a valid consideration.

- On the slower speed pump, the stuffing box will be larger in diameter, although possibly the same length. Since there is a gap between the packing and the shaft, the larger diameter packing would have a leakage rate proportionally larger than the smaller one.

- A negative aspect of the higher speed pump is the greater wear that may occur when pumping abrasive slurries. On the higher speed pump, the lesser surface of impeller reacting against the abrasive slurry will result in increased wear, although this may not be a sole consideration.

- Finally, the subject of noise generated by the pump may be in question. As a rule, the higher speed drive motor will be more noisy than the slower one. The motor noise can be reduced to a minor question by specifying a NEMA quiet standard, or by fitting large motors with air intake silencers. The subject of noise in the pumps themselves is extremely complex; it is fortunate that pumps are normally quiet machines. Noise level would be a criterion generally in areas such as hospitals, or offices where very low noise only can be tolerated. It should be noted that the physical size of the pump will have an effect on the generated noise. A noise shield on the pump and motor is generally effective, especially when a quiet pump is also specified.

DOUBLE- VS SINGLE-VOLUTE SELECTION

Pumps are normally designed to operate at or extremely close to 100 percent of design capacity and design head resulting in minimal radial thrust. In actual operation, however, load variances or miscalculation of the system head curve frequently result in pumping conditions changing either toward the "shut-off" or "run-out" condition. In such a situation, the resultant radial thrust on an impeller encased in a single volute can be great enough to cause premature wear and shaft failure. The threat is intensified because the force is cyclic in nature, due to rotation. Note points A and B on Fig. 17-1 graph.

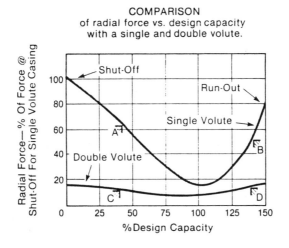

COMPARISON
of radial force vs. design capacity
with a single and double volute.

Fig. 17-1
(Source: PACO Pumps)

Traditionally, some manufacturers have attempted to deal with this condition through the use of larger, more expensive shafts and special bearings. PACO's innovative double volute, however, solves the problem by dividing the fluid flow into two similar geometric regions with two cut-waters 180° apart as shown in Fig. 17-2. Although the volute pressure inequalities remain as in a single volute casing, there are two resultant impeller radial forces opposing each

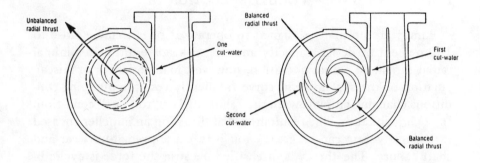

Fig. 17-2
(Source: PACO Pumps)

other. These opposing forces restrict the shaft to a normal axis of rotation, and the net radial force remains at a low level throughout the capacity range of the pump. See points C and D on the graph (Fig. 17-1).

In smaller pumps the radial forces on the shaft at the impeller are low enough that a single-volute design is normally adequate. As the impeller diameter and speed increase, however, unbalanced radial loads cause shaft deflection, resulting in premature wear of mechanical seals, bearings and wear rings.

PACO's compensated double-volute design effectively responds to this problem because the two volute areas create offsetting radial forces on the shaft. As a result, shaft deflection is minimized, operation is quieter, maintenance is reduced and pump life extended.

A double-volute pump should be specified for pumps in the specific speed range indicated in Fig. 17-3. The parameters shown on the chart are the result of extensive analyses by PACO engineers.

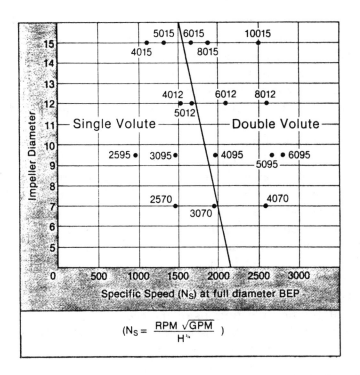

Fig. 17-3. A Double-Volute Pump should be specified when the specific speed falls to the right of the line in the chart.

(Source: PACO Pumps)

VARIABLE SPEED PUMPING

One of the best known facts in the world of pump people is that all pumps are oversized in both capacity and head. In some cases this is a safety factor; in many cases it is because the pump must be sized for the worst condition, which is seldom or never attained. Whatever the reason, it results in excessive wear and tear on the pump, consumes excessive power, and is deleterious to the valves and fittings in the pipe. Positive-displacement pumps handle this problem by using relief valves, which nevertheless incurs the problem of excessive power requirement as well as wear on the equipment.

Centrifugal pumps are a somewhat different problem—inasmuch as the oversize is hidden in the shifting of the pump curves, and throttling valves. The problems of excessive loadings on centrifugal pumps are more subtle than in the case of positive-displacement types. Radial and thrust bearing loadings, deflection of the shaft and seal wear, as well as excess power, are results of over or underloading the centrifugal pump.

In the not too distant past, steam-powered pumps were popular, because a turbine-driven pump could easily be driven at variable speed. It could be made to perform on the required curve. Then came cheaper power, with the result that almost all pumps were now electrically driven at constant speed. It was difficult to vary speed unless a mechanical speed-changing apparatus was inserted between the driver and the pump. Many such devices were used, but mechanical problems and poor efficiency minimized the use of such drives. Now, electric rates have risen to the point that even an expensive variable-speed installation will pay off in a very short time.

One of the best illustrations of the economy and usefulness of a variable-speed drive is obvious in the common boiler feed pump installation. In a steam boiler, the safety valves are set at some 10 percent above the highest operating pressure. Since the pump is expected to handle this pressure at full capacity, it is sized for excess pressure, above the boiler's needs. It is now obvious, that the boiler water level admission valve is designed to have a pressure drop of some 25 to 30 percent of the boiler operating pressure at full capacity. Now, since the boiler seldom operates at full capacity, the water admission valve will be required to further throttle the pump output. The pump will operate most of the time very far from its designed efficient curve. The wear and tear on the admission valve is also significant, making this equipment one of the plant high-maintenance items. Finally, to protect the pump from overheating at low flow, a low-flow bypass is provided for all proper installations, further adding to the unnecessary pump load. It is almost inconceivable that any boiler feed pump system should be installed for constant speed operation!

Of course, two-speed motors are sometimes specified, which means that neither speed will be correct.

Infinite variation is needed on systems where conditions of service are frequently or continuously changing. Applications might include maintaining a constant pressure at a variable flow for water supply; process installations where variation of flow or pressure is required to suit the output demand; installations with automatic control in response to other variables.

The use of variable speed for capacity control can save considerable power over the use of a throttle valve. This is illustrated by Fig. 17-4, in which the two solid line curves represent pump brake horsepower at constant speed (throttled discharge), and at variable speed for a system in which the resistance to flow is due entirely to friction loss. In such a system, pump brake-horsepower is proportional to the cube of pump speed, while capacity is directly proportional to speed.

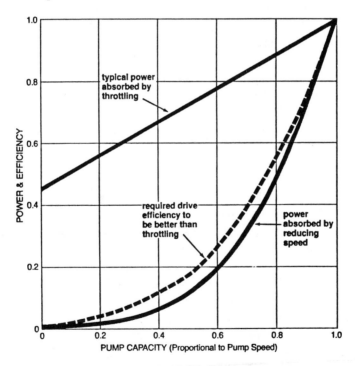

Fig. 17-4. Comparison of Capacity Control by Speed Variation and Throttline for Friction Head Loss System

The minimum efficiency required in a variable speed drive would be that which results in the same input power to the prime mover as for constant speed drive, and which can thus be represented as the ratio between the power at variable speed and the power at constant speed. This ratio, taken at various capacities, results in the dotted line in Fig. 17-4. Even high loss variable speed drives are considerably more efficient than this curve requires.

Because of the head/speed characteristic of a pump and the likely applications, a speed ratio of 5:1 will usually be the maximum required. This suits standard mechanical speed changers and is well within the range of electrical methods.

The importance of power factor of a motor or drive depends on whether the plant has facilities for power factor correction. If not, this should be included in the operating cost consideration.

In the "low loss" variable-speed drive, efficiency remains practically constant at all speed ratios and input power falls off with output power absorbed. This category covers drives such as thyristor-controlled dc motors and mechanical variators.

The "high loss" systems such as eddy current couplings and fluid couplings control speed by allowing a degree of slip between the input and output shafts while maintaining input shaft torque equal to output shaft torque. Because of the speed difference of the input and output shaft, theoretical maximum efficiency is equal to the speed ratio: for example, 50 percent at half speed.

The pump's cubic power/speed relationship helps in reducing losses. Fig. 17-5 shows the pump power absorbed and corresponding input to the coupling for a "high loss" drive. The losses, dissipated as heat in the coupling, are a maximum at .67 speed ratio of a value of 15 percent of full-load power. At speed ratios approaching 1, losses are quite small. In practice, of course, friction or resistance losses further reduce efficiency. Speed variation of induction motors by voltage control is essentially the same principal as introducing slip to reduce the speed. The resulting efficiency is similar to eddy current or fluid couplings.

For an optimum selection, drives must be considered in relation to the specific characteristics of the centrifugal pump involved, and of the system to which it is being applied. High specific speed pumps,

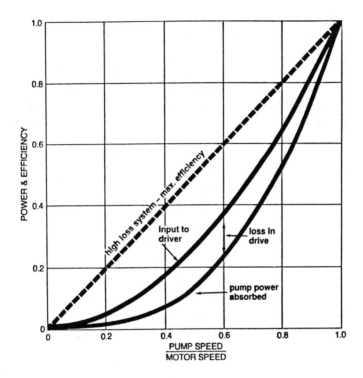

Fig. 17-5. High Loss Drive Characteristics Related to Centrifugal Pumps

for example, may require constant or even increasing horsepower as capacity is reduced by throttling, thus creating wider disparity between the two forms of power curves of Fig. 17-4. On the other hand, most pump system head requirements include static as well as friction components, under which circumstances the variable-speed horsepower curve does not come down to zero at zero capacity, and the spread between the curves is therefore diminished. Variations of this nature will affect the economics of variable-speed pumping.

A centrifugal pump has a cubic power/speed relationship, unlike most other types of industrial equipment. This characteristic can be used to reduce the size or complexity of the drive. For a centrifugal pump, torque is proportional to speed squared; power is proportional to speed cubed. The cubic falloff in power with speed is particularly beneficial in variable-speed motors, in which the cooling from a direct-coupled fan only falls off approximately as the square of speed; thus the need for forced cooling of the variable-speed drive may be eliminated. The power characteristic also means that high drive efficiency at reduced speeds is less important for a pump than for many other machines. If running time at reduced speed is only a small proportion of the total, power loss at reduced speed is minimal and low drive efficiency at low speed is not important. Nevertheless, it is important to know the operating speed or speeds you'll most often need in order to select a drive with the optimum efficiency/speed relationship. For example, a pump running at half capacity (half speed) with only occasional demands for full speed will have a motor eight times oversized for its usual service. Even if a drive of 100 percent efficiency is installed between a pump and induction motor, the overall efficiency of the set will fall at reduced speed because of the decrease in efficiency of the drive motor when operated at reduced loads.

The price per horsepower of a pump set is generally low compared to that for most other industrial equipment, with the drive representing a large proportion of the total. Variable-speed control makes the drive arrangement even more costly. A fixed change of speed, such as provided by a vee-belt, increases the combined pump and driver cost by a factor ranging from 1.2 to 2.4; variable speed can increase it by a factor from 2 to 11! Obviously, specifying more than you need can be costly, so a careful definition of drive requirements should be made to determine the real need for variable drive and the most cost-effective method of achieving it.

If a need for variable speed can be firmly established, however, it is helpful to remember that operating costs usually exceed the purchase price of a standard pump and motor in about six months.

The curve of Fig. 17-6 makes note of the fact that system efficiency, not necessarily pump efficiency, is the primary factor.

It is important to illustrate the relationship between pump power draw and operating cost. Given an electric motor drive operating at about 85 percent efficiency, each pump shaft bhp will draw about

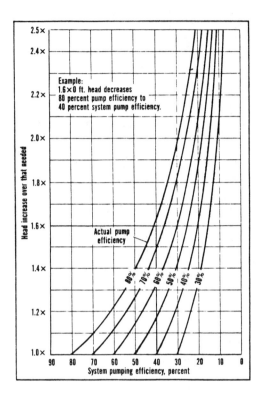

Fig. 17-6

0.88 kW. Thus, 10 pump shaft bhp will draw 8.8 kW. At a utility cost of $0.10 per kWh, the yearly cost for 10 bhp draw will be (10 bhp)(0.88kW per bhp)(24 hr per day)(365 days per yr)($0.10 per kWh)=$7709 per year.

If we can reduce power consumption by 10 bhp, we will save $7709 per year at $0.10 per kWh, $15,418 per year for a 20 bhp reduction, etc. The savings are substantial and, fortunately, can often be easily realized for oversized pumps if we establish and understand the basic physical correlations among pump power draw, the system flow-head relationship, circuit flow balance, and terminal heat transfer-flow relationships.

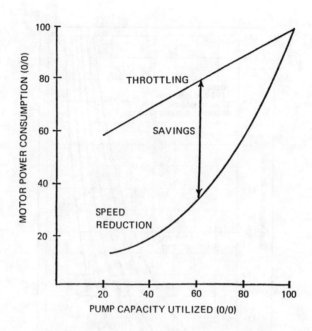

Fig. 17-7. Power Consumption vs Capacity Utilized for Conventional Pump Applications

The pump shaft brake horsepower requirement is determined by dividing the amount of power a pump puts into the water by the pump efficiency. The correlation is as follows:

$$bhp = whp/Ep = H\,Q\,p\,Php/33{,}000\,Ep$$

where

bhp	=	pump brake horsepower
Whp	=	water horsepower
Ep	=	pump efficiency
H	=	pump head, ft
Q	=	flow rate, gpm
p	=	density, lb per gal
Php	=	Horsepower, 33,000 ft-lb per min

Most pump horsepower curves assume an 85°F water density

(8.33 lb per gal). For this specific condition, the following relationship exists:

$$bhp = HQ/3960 \ Ep$$

Since many specifiers are interested in efficiency, they may become mesmerized by the term "pump efficiency" and neglect the terms in the pump power draw formula where true energy savings can be found (H×Q).

Pump power draw will increase directly with increases in either head or flow. The latter, however, is usually fixed. Therefore the only possible source of energy savings is by decreasing pumping head.

In addition to energy savings as well as reduced wear, with variable speed, there are a number of not-so-intangible benefits:

• It eliminates the present upper limit of 3600 or 3000 rpm imposed on squirrel cage 50 and 60 Hz motors. It permits the operation of pump and motor at 110 to 115 percent of rated speed.

• It improves the selection of specific speeds. It usually follows that increased speed means less initial cost and smaller equipment.

• Any appropriate design speed can be accommodated. There is no necessity to stay with synchronous speeds.

• It reduces the number of pump sizes that must be installed in large systems. A smaller number of sizes means less stocking problems for spares.

• Finally, a benefit which is not generally mentioned: the 10 or 15 percent extra speed available will compensate for possible errors in estimating the conditions of service.

VARIABLE SPEED CENTRIFUGAL PUMPING SAVES ENERGY*

Most process industries use centrifugal pumps to move liquids and slurries from point to point. In batch process, pumping is normally performed at a fixed speed. By contrast, continuous processes often require flow control; and it is here that a wise selection of the flow control means can yield substantial savings of energy.

---

*This section abstracted from material presented by Controlled Systems, Fairmont, W. VA.

Centrifugal pumps come in all shapes and sizes. They vary in capacity, in use and in basic design. Nevertheless, all centrifugal pumps, whether propeller, impeller, or turbine; low speed or high speed; single or multistage; have some characteristics in common. Because they work on the centrifugal principle, output flow varies directly with speed (at a fixed head), and output head varies as the square of speed (at a fixed flow). Consequently, the power to drive the pump varies as the cube of its speed.

At a given pump speed, the power required to drive the pump varies with the head. A fixed speed centrifugal pump will have a characteristic curve relating head and flow in a given medium (Fig. 17-8). The efficiency of the pump (fluid hp out divided by the shaft hp in) will vary, from zero at zero flow or zero head, to a maximum at about 60–75 percent of maximum flow.

When the pump speed is changed, the characteristic curve also changes. Fig. 17-8 shows a family of pump curves for varying speeds with the efficiencies shown. Note that as the pump flow is reduced at a fixed speed, there is a substantial reduction in efficiency. When the flow is reduced by changing speed, the reduction in efficiency is much less. How this affects energy savings may be deduced from Fig. 17-9.

When a pump is used to deliver a fluid to a process system, it usually encounters resistance to flow. This resistance has the general shape shown on the system curve in Fig. 17-9. It consists of a fixed value (the static head) plus a value which varies with flow (the friction head). When the pump of Fig. 17-8 is coupled to the system of Fig. 17-9, it will deliver a flow determined by the characteristic curves of the pump and of the system. This has been shown at the maximum useable flow. Interpolating from the data of Fig. 17-8, we find that the pumping efficiency at this maximum useable flow is 71 percent.

If the pump is to be used at some lower output, it is possible to restrict the flow by using a control valve or to reduce the speed of the pump.

The choice of the flow control means greatly affects the energy requirement of the pump. From Fig. 17-9, it can be seen that the pump efficiency, using a control valve, falls to 47 percent at 25 percent flow. Additionally, the system efficiency falls much faster,

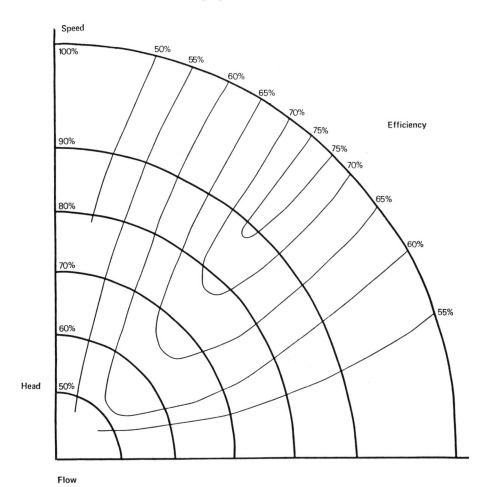

Fig. 17-8. Relationship Between Head, Flow, and Efficiency for a Typical Centrifugal Pump at Various Speeds

because a substantial amount of head energy is lost in the control valve. In the example, the head lost is equal to about 65 percent of the total, so that the actual efficiency of the pump and the control valve combined is only about 16.3 percent. By comparison, interpolating from the efficiency and flow curves in Fig. 17-8, the pump could deliver the required flow and head at about 60 percent of full speed at an efficiency of 59 percent.

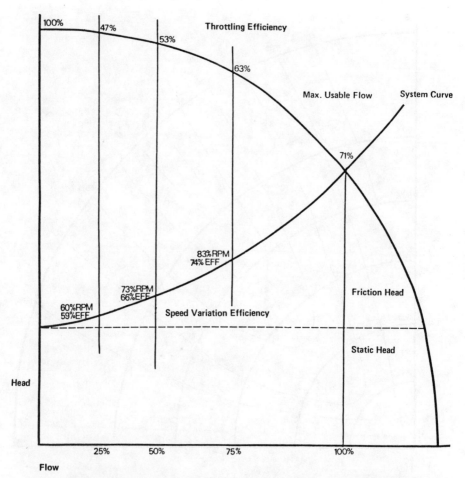

Fig. 17-9. System Curve with Alternative Pump Operating Points Showing the Choice Between Throttling and Varying Speed

The improved pump efficiency shown can be realized if the pump is driven by an efficient variable speed drive. Many variable speed systems, such as eddy current clutches or hydroviscous controls, are very inefficient at low speeds.

The efficiency of a 100 hp, squirrel cage, 3 percent slip, ac motor is about 90 percent at full speed and full torque, falling about 85 percent at full speed and one-quarter load. The same motor can also operate at half speed and half torque (one-quarter load) at about 90 percent efficiency. A good adjustable frequency drive, over the same range, will operate at 96 percent at full speed, full torque and about 93 percent at half speed, half torque.

An eddy current clutch has an efficiency of about 95 percent at full speed, full torque, but the efficiency falls sharply to about 47 percent at half speed, half torque. The efficiency of a motor driving the clutch is not affected by driving the clutch. However, the total efficiency is the product of the motor efficiency and clutch efficiency.

Table 17-1 gives a direct comparison of those three drives at loads equal to 25, 50, 75, and 100 percent of the system flow of Fig. 17-9.

Table 17-1. Drive Efficiency

Flow	AC Motor Only	AC Adjustable Frequency Includes Motor	AC Motor and Eddy Current Clutch
25%	85.0%	84.0%	49.5%
50	88.0	85.1	63.0
75	89.5	85.6	72.0
100	90.0	86.0	85.5

It is evident from the above figures that, in a time of rapidly increasing energy costs, the eddy current clutch is a poor choice as an adjustable speed drive.

When the drive efficiency is combined with pump efficiency, the overall efficiency can be computed. Table 17-2 compares overall efficiency of two systems—a variable speed pump with an adjustable frequency AC drive of good efficiency versus a fixed speed pump with a choke valve—at loads equal to 25, 50, 75 and 100 percent of the system flow of Fig. 17-9.

Table 17-2. Pump and Drive Efficiency

	Efficiency—Variable Speed			Efficiency—Fixed Speed + Throttle			
Flow	Pump	Drive*	System	Pump	Pump & Valve	Drive*	System
25%	59%	84.0%	49.6%	47.0%	16.3%	85.0%	13.9%
50	66	85.1	56.2	53.5	22.8	88.0	20.1
75	74	85.6	63.3	63.0	37.2	89.5	33.3
100	71	86.0	61.1	71.0	71.0	90.0	63.9

*Drive efficiency typical for motor in 60–100 hp range.

From the system curve of Fig. 17-9, it is also possible to calculate the relative amounts of energy, equivalent to head times flow, corresponding to each of the selected flow points. Using the system efficiencies for the two pumping schemes in Table 17-2, the relative electrical energy inputs can be computed for the two schemes. Table 17-3 shows those figures, which use the fixed-speed pump, operating wide open at 100 percent flow, as a base for comparison. The final two columns in Table 17-3 indicate the relative power savings available with an adjustable frequency drive compared to a fixed-speed motor. These are shown for the general case and also for a specific 100 hp pump, running 24 hours a day, in an area where power cost is a modest 6.0¢/kWh.

Of particular interest is the fact that in this example, the pump with a fixed-speed motor draws slightly more power at 75 percent than 100 percent flow. This can happen with some systems, but is not a universal occurrence.

The actual dollar savings will, of course, depend upon the pump characteristics, the horsepower rating, the system curve, the duty cycle and the cost of electrical power. The steps outlined in this section can be used in every case to project power savings accurately, and the savings will always be substantial.

Table 17-3. Relative and Real Savings

| | | Relative Electrical Input | | Reduction in Power Cost W/AC Variable Speed Drive | |
Flow	Relative Energy Output	Fixed Speed	AC Variable Speed	% of Full Power	For 100 HP Drive
25%	13.4%	61.8%	17.3%	44.5%	$19,386
50	31.4	100.0	35.7	64.3	28,014
75	57.4	110.3	57.9	52.4	22,830
100	100.0	100.0	104.7	(4.7)	(2,049)

If the terminals of a system, say a heat exchanger, are not balanced, the effect of increased head will be increased flow. Assuming that flow has increased to twice that specified, and that head is twice that needed the increased pump power draw over that required will be $(H \times 2)(Q \times 2)$ for a power increase of four times that needed. System pump efficiency will then become $75/4 = 18.75$ percent, even though the pump itself operates at 75 percent efficiency. Fig. 17-9 illustrates the choice between throttling and speed variation.

MULTIPLE PUMP PROGRAMMING WITH VARIABLE FREQUENCY DRIVE (VFD)[*]

With multiple pumps controlled in sequence by one VFD, a situation arises when a varied pump is switched to constant speed, or vice versa.

At each transition point a period of instability exists. With increasing effluent flow, a difference of 1 gpm or less in excess of the

---

[*]This section is abstracted from *Pollution Engineering*, August 1980.

maximum capacity of the operating pump(s) will require that a new pump be called upon and any operating pump(s) be switched to constant speed operation. The new variable-speed pump must then deliver the difference between the capacity of the previously operating pump or pumps and the influent rate. This difference could be 1 gpm or less.

In the case of a decreasing influent flow the situation is even worse. The influent rate must drop below the capacity of the pump(s) operating as constant-speed unit(s) before the variable-speed control is transferred and a pump turned off. Most centrifugal sewage pumps cannot be used in this type of operation without substantially increasing the probability of premature failure.

In a single-volute centrifugal pump, uniform or near-uniform pressures act on the impeller at design capacity (which coincides with best efficiency). At other capacities the pressures around the impeller are not even and there is a resultant radial reaction, Fig. 17-10. Fig. 17-11 shows a typical change in the radial thrust with changes in pumping rate. Specifically, radial thrust decreases from shutoff to design capacity and then increases with over-capacity. With over-capacity the reaction is roughly in the opposite direction from that with partial capacity. Note that the force is greatest at shutoff. The radial forces resulting from extreme low-flow operation can cause severe shaft deflection and ultimately shaft breakage with the danger becoming more severe with larger high-head pumps.

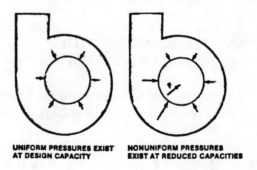

UNIFORM PRESSURES EXIST   NONUNIFORM PRESSURES
AT DESIGN CAPACITY   EXIST AT REDUCED CAPACITIES

Fig. 17-10

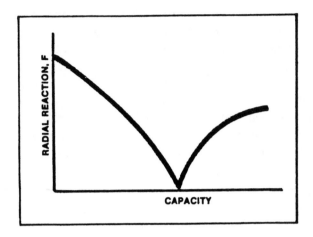

**Fig. 17-11. Magnitude of Radial Reaction in
Single-Volute Casing**

To overcome this potentially destructive problem as well as to
prevent sustained low-speed motor operation, most controllers intended for use in systems providing only a single controller for multiple pumps will include a minimum-speed adjustment. This prevents
sustained low-speed, low-flow operation, but causes cycling of the
lag or second lag pump at flow rates in excess of the capacity of any
pumps operating at constant speed but below the combined capacities of the constant-speed (C/S) operating pumps and the capacity
of the variable-speed (V/S) pump at its set minimum speed. A typical case would be:

Increasing or Decreasing Influent Flow

0 to 1000 gpm—lead pump (V/S)
1001 to 1250 gpm—lead pump (C/S), 1st lag (on/off at minimum
speed)
1250 to 2000 gpm—lead pump (C/S), 1st lag pump (V/S)
2001 to 2250 gpm—lead pump (C/S), 1st lag pump (C/S), 2nd lag
pump (on/off at minimum speed)
2250 to 3000 gpm—lead pump (C/S), 1st lag pump (C/S), 2nd lag
(V/S)

This cycle is less disruptive than that experienced with full-speed on/off controls but would still deviate from the ideal variable speed system. An additional consideration is the necessity for transfer contactors and the control circuitry necessary to switch the controller from one motor to another, Fig. 17-12. Such a control system is more complex and therefore more maintenance-prone than that required for a system having individual speed controllers. Also, under certain flow conditions the cycling of the lag pump could be fairly rapid, causing excessive wear on the transfer contactors.

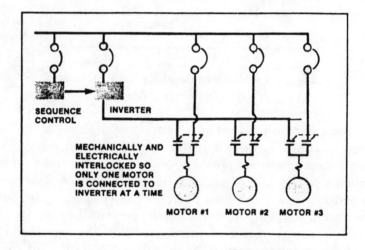

Fig. 17-12. Typical Three Pump VF One-Line Diagram

The fact that the everyday operation of a control system using one controller for multiple pumps is less than perfect may be acceptable in view of the cost savings. The operation in a failure mode is not. Any failure that disables the variable-speed controller produces an on/off constant-speed operation of the pumps. Although this is theoretically permissible with variable-frequency controls, satisfactory or even semi-satisfactory operation requires that the installation be designed with this alternative in mind.

In a constant-speed station the wet well must be sized to provide sufficient storage to limit the minimum cycle time to one which is acceptable to the equipment used. By contrast the wet well in a variable-speed pump station is sized only for proper hydraulic flow and access, and to assure a reasonable gain in the control range. (Gain is the ratio of a change in input to the resulting change in output and should not exceed a 5:1 ratio if a stable system is expected.) In terms of wet well size this means that the storage capacity of the control range can be as little as one-fifth the capacity of the pump or pumps controlled over that control range.

Heat and Wear

All equipment items involved in the starting process are subjected to severe stress during this period. Obviously starters and contactors have a mechanical life measurable in terms of a finite number of closures, with a substantial portion of this life being lost in one week of operation in a short-cycling mode.

Motor and pump bearings are subjected to their most severe strain during starting, and the pump itself must endure periods of stress from vibration and low-flow shaft deflection each time it is started or stopped. As a result the mechanical life of the pump and motor will be reduced significantly by any prolonged period of this short-cycle on/off operation.

In addition to the mechanical wear there are thermal constraints on the operation of any type of electrical equipment. Starters and contactors are accepted for rating at a given horsepower based on their ability to handle the current associated with that horsepower without over-heating. During the starting cycle a 3-phase induction motor may draw from 5 to 7 times normal current. Under the conditions imposed by an extremely rapid duty cycle, the current-carrying components of the starter may not have time to dissipate the additional heat created by this current surge before the motor must again start. Over a period of time the contacts may overheat causing their premature erosion. In extreme cases the contacts may even become welded together which will cause the motor to keep running and possibly damage the pump.

All thermal ratings given to motors are based on a certain temperature rise from a specific ambient temperature. The motor fan, cooling fins, and general construction for a given rating are designed so that the motor will not exceed the limiting temperature when operated continuously at rated load, speed, and voltage. A given motor by a specific manufacturer may have an additional safety factor built in, and occasionally engineers write specifications calling for Class F insulation and a Class B rating. This is simply an additional safety factor requiring that the frame size and fan be capable of maintaining the temperature rise to a low value, but that the insulation be capable of withstanding a higher temperature.

It can be readily seen that any questions regarding the capacity of a given motor to withstand repeated starts without thermal damage can be answered only by the motor manufacturer, and then only after a careful evaluation of all significant parameters. However, even with an exacting analysis by the motor manufacturer and the closest coordination with the controller manufacturer to secure the most suitable motor for the specific application, problems can occur. In those systems with one controller for two or more pumps, the effect of elevated temperature, instead of being immediate and catastrophic, will be to reduce the overall operating life of the motor. This could cause a motor designed to last 20 years to fail in 10 or 15.

Wound Rotor Motors with
Secondary Control

The wound rotor motor has been used with a variety of secondary controls from step resistors to electronic power recovery units. Theoretically any wound rotor motor is capable of operating at full speed with the slip rings shorted. Indeed this is customary in many applications where the reason for using this type of motor is its very favorable starting characteristics. A wound rotor motor, however, should never be started with the slip rings shorted. Without external resistance in the motor secondary circuit, damage to the slip rings and brushes is almost certain to occur.

When the motor is energized with the rings shorted, the entire initial current surge will pass through the brushes to the slip rings. This current, at 300 to 400 percent of normal full-load values, will result in burn spots on the rings. Frequently the starting surge will cause small particles of copper to be melted and deposited on the face of the brush; these then lead to threading or grooving on the slip ring surface. The life expectancy of a set of brushes in a properly controlled wound rotor motor is in the neighborhood of 3 to 5 years, with ring maintenance in the 5- to 7-year category. If the motor is started across the line, the life expectancy can be measured in hours or days at best, with larger motors being more susceptible to damage.

Summing It All Up

The cost-effectiveness of a control system using one controller for two or more pumps is highly questionable for the following reasons:
In normal operation:

1. It is inherently unstable at transition points.
2. There is a danger to the pumps if minimum flow rates are set too low.
3. The switching circuitry and contactors are sources of potential service problems.

ANALYSIS PROBLEMS*

Cost of a variable-frequency drive is so moderate and the energy-saving potential so great that many users feel that there is no need for detailed analysis of true saving or operational and maintenance gains. While this may be true for many installations, some significant problems have been encountered as experience is gained in the application of these drives. The problems detailed here should be fully understood by anyone evaluating a variable-frequency-drive installation.

Use Affinity Laws Correctly

The energy saving obtained by driving a pump at reduced speed is found by locating a new head/flow curve that intersects the re-

---

*This section is abstracted from *Power*, February 1987.

quired new operating point (Fig. 17-13). Head and flow at the new operating point are given and it is necessary to find the new speed and horsepower. The new horsepower is then compared with the full-speed horsepower to find the energy saved. The problem arises from the incorrect use of the affinity laws to locate the new operating point. Affinity law (5) in Fig. 17-14 (derived from equations 2 and 3), states that head is proportional to the square of flow, and can be plotted as an affinity curve as shown in Fig. 17-13.

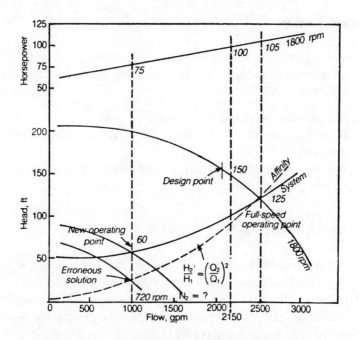

Fig. 17-13. Full-Speed Operating Point is Incorrect Initial Condition for Affinity-Law Solution of New Operating Conditions

A common mistake is to take either the pump full-speed operating point or the design point as the initial condition and assume that the pump operating point follows the affinity curve from there to the new operating point as the speed is reduced. This is wrong because, when the pump speed is changed, all the variables change. In effect this

$$hp = \frac{H \times Q \times SG}{3960 \times E} \qquad (1)$$

$$\frac{Q_1}{Q_2} = \frac{N_1}{N_2} \qquad (2)$$

$$\frac{H_1}{H_2} = \left(\frac{N_1}{N_2}\right)^2 \qquad (3)$$

$$\frac{hp_1}{hp_2} = \left(\frac{N_1}{N_2}\right)^3 \qquad (4)$$

$$H_2 = H_1 \left(\frac{Q_2}{Q_1}\right)^2 \qquad (5)$$

Where hp = Horsepower
H = Head, ft
Q = Flow, gpm
SG = Specific gravity
E = Efficiency (fraction)
N = Speed, rpm

Fig. 17-14. Horsepower and Affinity Equation Define Relationships Between Operating Conditions of a Fan or Pump

is an attempt to find three variables (flow, head, and pump speed) with only two defining equations. The affinity curve shown on Fig. 17-13 will locate a new operating point, and probably predict significant energy saving. But when the head is checked using the affinity equation (3), it is found to be too low. This means that the horsepower savings predicted are unrealistic. An example of this erroneous calculating procedure is shown in Fig. 17-15.

Another method sometimes used to find the new operating point is to calculate the efficiency at full-speed and assume that it is the same at the new operating point. It's then possible to calculate

Initial Conditions:
$Q_1 = 2500$ gpm
$H_1 = 125$ ft
$hp_1 = 105$
$N_1 = 1800$ rpm
Find new speed using equation (2)

Final Conditions:
$Q_2 = 1000$ gpm
$H_2 = 60$ ft

$$N_2 = N_1 \frac{Q_2}{Q_1} = 1800 \frac{1000}{2500} = 720 \text{ rpm}$$

Find new horsepower using equation (4).

$$hp_2 = hp_1 \left(\frac{N_2}{N_1}\right)^3 = 105 \left(\frac{720}{1800}\right)^3 = 6.7 \text{ hp}$$

This is a large decrease in horsepower (horsepower would be 75 for 1000 gpm at 1800 rpm, Fig. 17-13). Verify its accuracy by checking the final head using equation (3).

$$H_2 = H_1 \left(\frac{N_2}{N_1}\right)^2 = 125 \left(\frac{720}{1800}\right)^2 = 20 \text{ ft}$$

Since the system requires 60 ft head at 1000 gpm flow, the calculated speed is obviously too slow. The solution found by this approach is shown on Fig. 17-13. It can be seen that it is improper since the derived head/flow curve does not intersect the desired new operating point.

Fig. 17-15. Incorrect Value of Horsepower at Reduced Speed Results Here Because Full-Speed Operating Point is Used as Initial Condition

the horsepower at the new operating point and find the energy saved. But once again, when the head or flow at the new operating point is checked it is found to be too low to meet the system curve. Incorrect choice of initial conditions is again the reason.

Use Iteration or Graph Curves

The correct initial condition, from which the affinity laws can be used to find the new operating point, lies somewhere along the pump characteristic, but is neither the full-speed operating point or the design point. To find it using iteration, choose a new operating speed (N_2). Then use the affinity equations and the pump curve of Fig. 17-13 to find the flow and head at this speed. Then use affinity equation (3) to see if the head at the new condition satisfied the system curve. If it is too low, repeat with a higher initial speed, if too high, try a lower initial speed.

The same thing can be done graphically (Fig. 17-16). Construct an affinity curve that passes through the origin and the new operating point, using the known square relationship of equation (5). The point at which the affinity curve intersects the pump's head/flow is the correct initial condition. The new speed is then found from equation (2) or (3). The new horsepower can be found using this speed; the initial horsepower, as read from the graph; and equation (4).

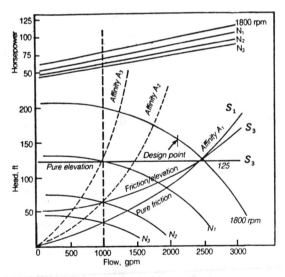

Fig. 17-16. Graphic Method Finds Correct Initial Conditions on Pump Characteristic Head/Flow Curve

Is Head Friction or Elevation?

Pumps can encounter three types of system curves: pure friction, pure elevation, and combination friction/elevation. Each generates a different system curve, (S_1, S_2, S_3) as shown in Fig. 17-17. Suppose it is necessary to find the speed and horsepower to deliver 1000 gpm to each of these systems. Final conditions are dictated by the required flow and the pressure needed to deliver it in each case. The correct initial conditions can be found by plotting affinity curves (A_1, A_2, A_3) (Fig. 17-16) for each system. These curves pass through the origin and the head required by each system at 1000 gpm (Fig. 17-17). They reveal some very important points.

First, only the system curve for the pure-friction case coincides with the affinity curve. This is the only situation in which the full-speed operating point can be used as the initial condition. For pumps it is very rare. Virtually all pumps work against some elevation.

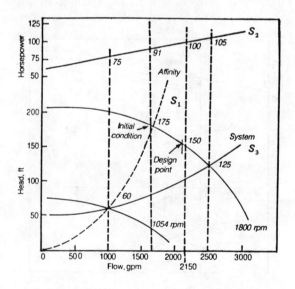

Fig. 17-17. System and Affinity Curves are Different for
Pure Friction, Pure Elevation, and Combined
Friction/Elevation Systems

Fig. 17-17 also shows that the potential energy saving with variable-frequency drives are highest with pure-friction systems and lowest with pure-elevation systems. Although most systems are a combination of friction and elevation, they are often operated as pure-elevation systems. This is particularly true for a parallel-pump system serving multiple uses. Usually, some of these users require a certain minimum head. In the absence of variable-speed operation, an operator must start and stop pumps as needed to ensure that the header pressure is at least as high as this minimum.

With variable-speed operation, one or more pumps is varied in speed to maintain the minimum header pressure. Thus each pump sees a constant-head system and energy saving must be calculated on this basis. If horsepower is calculated on the basis of a pure friction (which has been done) the saving will be grossly over-stated.

Remember, also, that real pumps diverge from the ideal behavior defined by the affinity equations. Errors resulting from the use of these equations increase as the speed and size of the pump decreases. Many pump manufacturers offer curves that give this deviation as a function of speed and pump size.

Need Computer to Find Saving

The manual iterative and graphical methods of finding the energy saving are combersome; also, they only find the horsepower at one operating point. Real system conditions vary widely and sufficient operating points should be analyzed to be representative of future demand. Since economic evaluations are based on annual cash flow, enough calculations must be made to be representative of variability in typical future years. These extensive calculations are best handled by a computer.

Thorough analysis requires extensive data on the pumps and system, including the following:
- Flow/head curve at a fixed running speed
- Horsepower/flow (or efficiency/flow) curves for the fixed speed
- Electric-drive-motor efficiency curve
- Variable-frequency-drive efficiency curve
- Minimum allowable stable flow

- Flow/head curve for system
- Flow demand/time curve.

Computer spreadsheet programs are very helpful for the analysis. First convert the pump, motor, drive, and system curves to workable equations using regression. Then write an algorithm to calculate the saving using the methods given here. Repeat this computation as many times as necessary to represent future usage. A very clear picture of potential saving is possible by analyzing past demand data on an hourly or daily basis.

Check Instability, Vibration

All centrifugal pumps require some minimum flow for stable operation. Below this minimum, recirculation, cavitation, and unbalanced radial forces occur. The result is severe vibration that dramatically increases maintenance and may even destroy the pump.

Clearly a variable-speed drive may put the pump into areas of crtical vibration. These may be within the normal operating speed range or in overspeed conditions. Careful consideration must be given to these possibilities before the pump is operated. Minimum-flow needs can be obtained or calculated from vendor's data or other literature. Expected critical speeds can be calculated or determined by test.

Special controls are needed to avoid damage to the pump because of low-flow instability. Normally, the variable-frequency drive changes speed in response to a controller that is controlling temperature, pressure, flow, or level. To protect the pump, an independent flow-control loop is needed, incorporating flowmeter, transmitter, and controller. This is in addition to the primary-control loop and is arranged to override the primary variable during low flow. To keep the primary variable under control, a control valve or bypass loop is necessary.

A common misconception is that the minimum flow criterion can be met by limiting the drive to a minimum speed. This is only true if the pump head can be limited to an exact maximum value at minimum flow. This may be possible with a single pump serving a single user, but this situation is rare.

The final control range of a variable-speed pump should be based on actual vibratory conditions and motor- or drive-current limits. Determine minimum stable flow by experimentation. Once the onset of instability is determined, set minimum flow slightly above this. Base the upper speed limit on motor or drive current. Finally, explore the entire range for any significant criticals and, if any are found, set the drive to avoid running at these speeds.

Key Economic Factors

Experience is proving that capital investment and operating costs for variable-frequency installations may be higher than anticipated. Here are some critical areas:

Minimum flow requirements reduce the range over which energy saving can be achieved. Too often, engineers do not consider this and thus produce overly optimistic results. Also, the rather extensive control systems needed to protect the pump from instability, add to the initial capital cost.

Maintenance cost savings projected for variable-frequency drives are usually based on the elimination of control valves and reduced wear on the pump, because it is running at lower speeds. However, control valves are frequently needed to maintain stability and it is unlikely that they can be eliminated, particularly in a retrofit application. It is true that the pump will run at a lower average speed, but this fact is often offset by the fact that the pump is called upon to operate on the borderline of its stable design and, sometimes, at flows beyond its design speed.

The variable-frequency drive is a complex piece of equipment that needs periodic maintenance and will fail on occasions. Since there will probably be no pump or control-valve maintenance saving, it is prudent to decrease the projected saving by the expected maintenance cost of the drive.

Air-conditioning system may be needed to prevent malfunction of the drive electronics. This represents additional capital cost.

Negative saving occurs whenever the variable-speed pump is required to operate at or near its full speed. This is because of some inefficiency in the variable-frequency drive. Too often these periods of increased operating cost are not included in the evaluation.

Oversize systems produce very large projected cost savings for the variable-frequency drive. Often these systems were originally made larger than necessary because of conservative design information or expected future expansion. If it can be shown that the capacity is not needed, it is often more cost-effective to reduce the capacity rather than install a variable-speed drive. This may be done by installing a smaller impeller or a motor with a lower speed and horsepower.

BOILER FEED PUMP INSTALLATION

Earlier in this chapter, it was stated that a boiler feed pump installation was among the best areas for the use of the variable-frequency drive. There are several ways to arrange such a system. In the increasing order of sophistication:

1. *One pump, one boiler, no feedwater regulating valve.* The savings are obvious: deletion of valve installation and maintenance cost; minimum head requirements. The pump speed is varied according to the level of water in the boiler. The level control system used for the feedwater admission valve transmits its signal directly to the pump VFD controller. (Fig. 17-18(a))

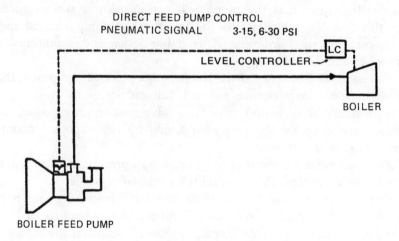

Fig. 17-18(a)

With the advent of single-boiler systems, a new era of boiler feed pump control has been established. With single-boiler propulsion plants, it is possible to eliminate not only the feed pump constant discharge or constant differential pressure controller but also the boiler feedwater regulating control valve and, thereby, cut initial capital investment. It should also be noted that by eliminating the boiler feedwater control regulating valve an inherent efficiency loss due to throttling of feedwater input to the boiler is eliminated.

2. *Constant discharge pressure control.* (Fig. 17-18(b)) The feed pump is controlled to a predetermined pressure setting irrespective of plant load. The advantage of this system is that the pump will not be required to operate at near shut-off pressures, due to the shifting of the operating point on the curve. This system is not as good as (3), and will cost as much.

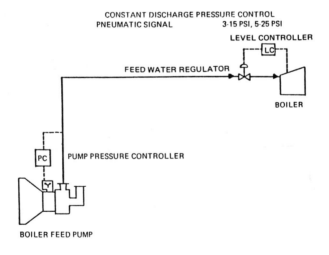

Fig. 17-18(b)

3. *Constant differential pressure control.* Feed pump pressure is controlled to produce a predetermined pressure drop across the feedwater regulating valve, usually approximately 50 to 75 psi (3.5 to 5.5 atmospheres) thus allowing the boiler feed pump to follow plant demand.

The latter system provides a more efficient control in that the feed pump output capabilities vary as plant demand varies. Note that the feedwater pump should be capable of providing about 10 percent excess pressure, over boiler pressure, to satisfy feedwater requirements when all safety valves are discharging. In mode (3), the feedwater pump discharge pressure is automatically raised if required when safety valves are open. The excess pressure is not required for normal pressure. Fig. 17-18(b) reflects a typical system schematic diagram for constant discharge pressure control and Fig. 17-18(c) reflects a constant differential pressure control system. There is, however, a loss in the differential pressure across the feedwater regulator.

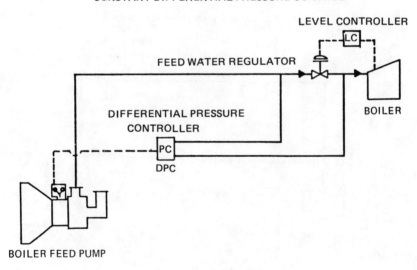

CONSTANT DIFFERENTIAL PRESSURE CONTROL

LEVEL CONTROLLER

FEED WATER REGULATOR

DIFFERENTIAL PRESSURE
CONTROLLER

BOILER

DPC

BOILER FEED PUMP

Fig. 17-18(c)

Note that system (3) can be made even more desirable by adding a load correction to the differential control. The less the load, the less the differential. See Fig. 17-18(d).

VARIABLE DIFFERENTIAL PRESSURE CONTROL

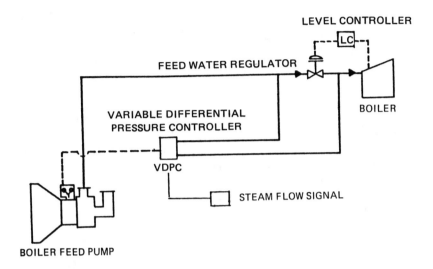

Fig. 17-18(d)

Note that an additional benefit of this system is that at low loads, the feedwater valve will open wider, thus reducing wear.

A more detailed block diagram for various arrangements of variable-speed control is shown by the Bell & Gossett data sheets, Fig. 17-19(a) and (b).

METHODS FOR VARIABLE-SPEED PUMP CURVE PLOTTING*

Variable-speed pump curves are invaluable to analyze and solve hydronic system pumping dynamics where variable-speed pumping is to be considered.

---

*This section courtesy of ITT Fluid Technology Corporation, 1984.

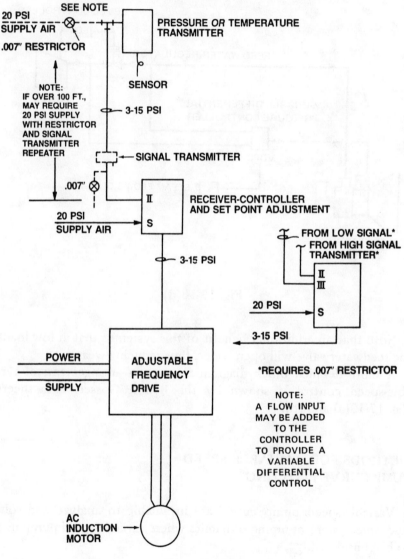

DIFFERENTIAL PRESSURE OR TEMPERATURE PNEUMATIC CONTROL BLOCK DIAGRAM

Fig. 17-19(a)

(ITT Fluid Technology Data Sheet B-841)

ELECTRIC PRESSURE CONTROL BLOCK DIAGRAM

ELECTRIC TEMPERATURE CONTROL BLOCK DIAGRAM

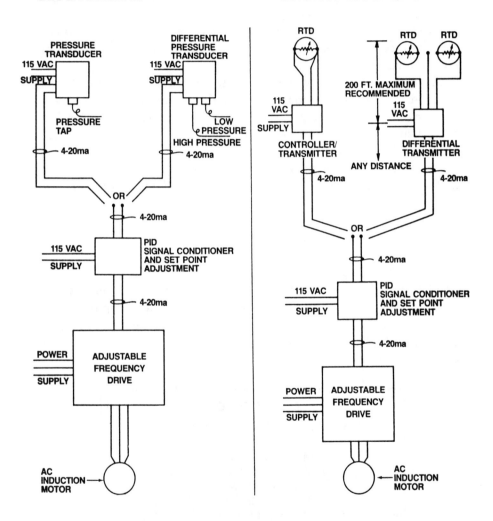

Fig. 17-19(b)

(ITT Fluid Technology Data Sheet B-841)

Currently, all of ITT Bell and Gossett's published pump performance curves are at constant speed based on the number of poles and frequency of the motors (i.e., 60 cycle=3500, 1750, 1150: 50 cycle=2900, 1450 rpm, etc.).

Until such time that accurate, computer-generated, variable-speed performance curves become available, approximate curves should be constructed and used.

Since the curves are usually used to determine operating power levels at less than design load (gpm) by superimposing system curves, the accuracy of the pump curves are not too critical since other factors such as piping pressure losses and load-demand profiles are based on "best available" estimates.

The methods following are in order of accuracy. Selection of the method may be limited by the availability of data to construct the curves.

Method #1 (Fig. 17-20)

From available published curves for 3500, 2900, 1750, 1450, and 1150 rpm speeds, replot the desired diameter curves on a single graph paper. Limitation: Due to differences of impeller bores between the 1750 and 3500 rpm pumps for sizes larger than 2AC, 3500 and 2900 rpm curves should not be plotted on the same sheet with the 1750, 1450, and 1150 rpm curves.

DATA SYMBOLS

R_1 = RPM @ known condition #1
R_x = RPM @ desired condition X
G_1 = GPM @ known condition #1
G_x = GPM @ desired condition X
H_1 = TDH @ known condition #1
H_x = TDH @ desired condition X
E_1 = Efficiency @ known condition #1
E_x = Efficiency @ desired condition X
HP_1 = Horsepower @ known condition #1
HP_x = Horsepower @ desired condition X

Plot variable-speed curves as follows:
1. Replot curves for the desired diameter from each available speed.
2. Locate numerically equal efficiency points along their respective speed curves.
3. Connect equal efficiency points with a smooth curve.
4. Intermediate speed curves may be estimated by starting the curve from the 0 gpm scale and constructing it parallel to the nearest existing speed curve.
   The starting point head, H_X is:

$$H_X = \left(\frac{R_X}{R_1}\right)^2 H_1$$

Alternate Method: Use linear extrapolation if the curves (speeds) are spaced relatively close together; such as between 1450 and 1750 rpm.

Method #2 (Figs. 17-21 and -22)

1. Replot performance curve for desired impeller diameter at 60 Hz (full speed) onto a blank graph paper.
   Alternate: Use copy of existing constant-speed curve and superimpose plots on it.
2. Plot affinity (constant efficiency) curves through desired efficiency points located along the full speed curve.

$$H_X = \left(\frac{G_X}{K_1}\right)^2 \qquad\qquad K_1 = \left(\frac{G_1}{(H_1)}\right)^{1/2}$$

3. Calculate G_X and H_X along each efficiency curve for the desired rpm, R_X. Use calculation form.
4. Plot G_X or H_X from Step 3 along the constant efficiency curve for each rpm, R_X.
   Note: Locate either G_X or H_X on the efficiency curve, selecting the value which can be read most accurately along the H (TDH) or G (gpm) scale. The other value may be used for cross-checking.
5. Connect all of the equal rpm points on the constant-efficiency

curves with a smooth curve. Note that the rpm curves are nearly parallel to the constant-speed impeller-size curves.
6. Intermediate rpm may be located along any efficiency curve by referring to a known G, H and R.

$$G_X = \left(\frac{R_X}{R_1}\right) G_1 \qquad \text{or} \qquad H_X = \left(\frac{R_X}{R_1}\right)^2 H_1$$

However, for best accuracy, refer to G and H at full-speed conditions.
7. Horsepower at any point may be calculated from

$$HP_X = \frac{G_X \times H_X}{3960 \times E_X}$$

Method #3 (Fig. 17-23)

For a faster, though less accurate method of plotting the variable-speed curves, the existing impeller diameter curves may be converted to rpm curves as follows:
1. Plot affinity (constant efficiency) curves as described in #2-1 and #2-2 from the *existing constant-speed* pump curve. Two or three will suffice at this time.
2. The *impeller diameter* curves may be relabeled as specific rpm curves by relating G_X or H_X at the intersection of any affinity curve with the impeller diameter curve and calculating the equivalent speed R_X:

$$R_X = \left(\frac{G_X}{G_1}\right) R_1 \qquad \text{or} \qquad R_X = \left(\frac{H_X}{H_1}\right)^{1/2} R_1$$

Where G_1 and H_1 are values at the intersection of the *same* affinity curve (from which G_X or H_X was read) with the desired impeller diameter curve at full speed R_1.
Since the accuracy of calculating the unknown rpm, R_X, depends on the accuracy of reading G_X, H_X, G_1 and H_1, several points along the impeller diameter curve should be "tested" for R_X and the average value used.

Example: 1510-4E-1750 with 1½" impeller. Determine the rpm for the 9½" diameter curve.

Step 1—Plot affinity curve passing through the intersection of one of the efficiency curves and the 10½" curve. Say the efficiency curve is for 70 percent.

Step 2—Read values G_1 and/or H_1 at the point of intersection between the affinity curve and 10½" curve: $G_1 = 330$, $H_1 -$ 104.5, $R_1 = 1750$.

Step 3—Read values G_X and/or H_X at the point of intersection between the affinity curve and 9½" curve: $G_X = 300$, $H_X = 85$, $R_X = ?$

Step 4—Calculate R_X from gpm and head.

$$R_X = \left(\frac{300}{300}\right) \times 1750 = 1591 \text{ rpm}$$

$$R_X = \left(\frac{85}{104.5}\right)^{1/2} \times 1750 = 1578 \text{ rpm}$$

The above averages to about 1585 rpm for the 9½" curve. If above R_X variance is too large, plot another affinity curve for a different efficiency value and compare R_X values with above.

Step 5—To calculate hp at any speed, plot or estimate the efficiency at the point in question remembering that the efficiency remains constant along the affinity curve.

Note: Disregard the hp curves from the original pump curves as they are "full speed" values. Calculate hp from expression shown in method #2, step 7.

See Figs. 17-20 through -23 for curve plots of Methods #1 through #3.

VARIABLE SPEED PUMP CURVE PLOT
METHOD #1

1510 2AC—6½″ IMP.

65%

CONNECT EQUAL EFF. POINTS

3500 RPM

2900

$$H_v = \left(\frac{2400}{3500}\right)^2 164 = 77$$

$$HP = \frac{65 \times 110}{3960 \times .65} = 2.78$$

2400

1750

1450

TOTAL HEAD IN FEET

CAPACITY IN U.S. GALLONS PER MINUTE

Fig. 17-20

VARIABLE SPEED PUMP CURVE PLOT
METHOD #2 (REPLOTTED)

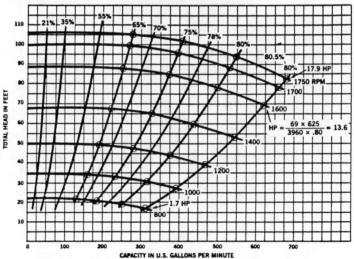

21% 35% 55% 65% 70% 75% 78% 80% 80.5% 80% 17.9 HP

1750 RPM
1700
1600

$$HP = \frac{69 \times 625}{3960 \times .80} = 13.6$$

1400

1200

1000

1.7 HP

800

TOTAL HEAD IN FEET

CAPACITY IN U.S. GALLONS PER MINUTE

Fig. 17-21

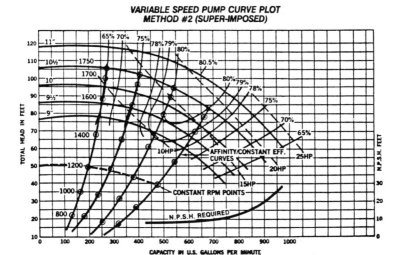

Fig. 17-22

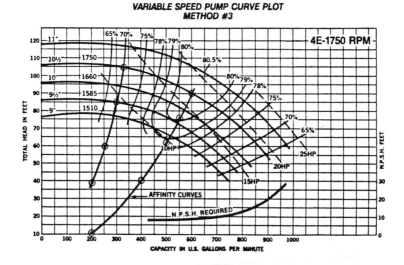

Fig. 17-23

CHANGING PUMP SPEEDS

A quick method for calculating the performance of turbine, propeller or mixed-flow pumps at any speed is given below. All that is required is a slide rule with A, B, C, and D scales and a performance curve of the pump at any speed. (*Note:* The "A" and "B" scales on the slide rule do not have to be on the same side of the rule as the "C" and "D" scales, but it is more convenient if they are.)

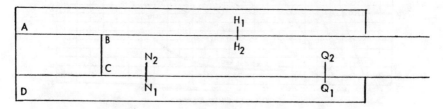

N_1 = rpm
H_1 = head }Pump performance on published curve.
Q_1 = capacity

N_2 = rpm
H_2 = head }Pump performance at desired speed.
Q_2 = capacity

1. Set desired pump rpm (N_2) on the "C" scale over known pump rpm (N_1) on the "D" scale. *DO NOT MOVE SLIDE FROM THIS POSITION.*

2. Select any point on the published curve (H_1, Q_1). When these values are located on the "A" and "D" scales as shown in the diagram, the performance (H_2, Q_2) at the desired speed can be read directly from the "B" and "C" scales. The efficiency at this new point (H_2, Q_2) may not be the same as the efficiency on the published curve at (H_1, Q_1), and therefore there is some small error. Refer to Fig. 17-24, which shows the efficiency loss predicted for pumps at reduced speed. These curves, of course, are very general. Actual efficiencies vary based on size, pressure, speed, clearances and many other design parameters.

A complete new curve can be constructed by selecting additional points.

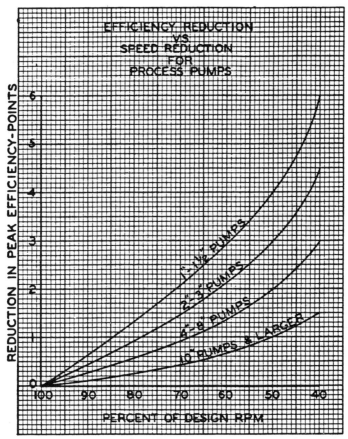

Fig. 17-24

ADJUSTABLE-FREQUENCY DRIVE (AFD) ARRANGEMENTS

Adjustable speed pumps controlled and powered by an Adjustable Frequency Drive (AFD) can be set up to respond to level, flow, temperature, or pressure parameters. One pump may be used or several. Usually, because of the cost of the AFD, only one control unit may be used to control up to four pumps. More than that number, the switching becomes cumbersome. Consider a set of pumps being operated to control the level in a wet well.

Adjustable speed is obtained by converting fixed-frequency, a-c power to adjustable d-c voltage, which is then inverted to adjustable-frequency a-c power. Motor speed is proportional to the adjustable-frequency a-c power supplied from the inverter. The basic elements of a typical inverter drive are shown in the simplified diagram (Fig. 17-25). Three-phase a-c power enters the power unit through the current-limiting fuses and is fed to the three-phase, full-wave, phase-controlled converter. The resulting d-c bus voltage is adjusted to a value required by the a-c motor. After suitable filtering, the variable d-c voltage is delivered to three identical inverter phase modules. Each of these modules can supply current in either direction from the d-c bus and can be switched "on" or "off" at high speeds. The inverter control establishes the fundamental frequency at which the voltage will be supplied to the a-c motor.

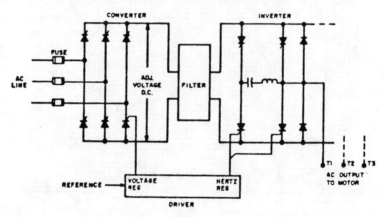

Fig. 17-25. Simplified Inverter Diagram

The reference enters the driver regulator and is channeled through the appropriate logic elements to provide firing signals to the converter and inverter modules. The converter's d-c voltage and the inverter's fundamental frequency are adjusted directly proportional to the reference producing a constant volts per-Hertz output to the a-c motor.

Simplex

The simplex system consists of one adjustable speed pump controlled by one adjustable-frequency inverter. The motor operates at the desired speed as determined by the speed potentiometer in the manual mode or the external process signal in the automatic mode.

Duplex

The duplex system consists of two adjustable speed pumps (lead-lag) controlled by one adjustable-frequency inverter. On increasing flow, the lead motor is connected to the inverter and accelerated from zero to the preset minimum speed. The motor speed is then automatically varied between minimum and maximum speed in response to a proportional signal from the transducer.

If the flow (for instance into the wet well) increases beyond the capacity of the lead pump operating alone at maximum speed, the lead motor is transferred by a closed transition function from the inverter to the fixed-frequency a-c line to run at constant speed. The lag motor is then connected to the inverter and run at an adjustable speed as determined by the level in the wet well.

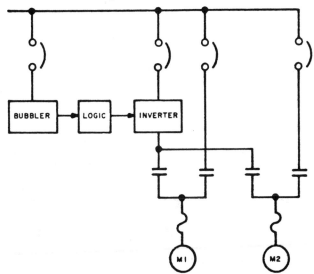

Fig. 17-26. Duplex One-Line

As the wet well level decreases, the lag motor is decelerated to a preset minimum speed. As the level continues to decrease, the lag motor will be disconnected from the inverter and stopped. The lead motor is then connected to the inverter by a closed transition function and run at an adjustable speed as determined by the level in the wet well.

Triplex

The triplex system consists of three adjustable speed pumps (lead, lag 1, and lag 2) controlled by one adjustable-frequency inverter. On increasing flow, the lead motor is connected to the inverter and accelerated from zero to the preset minimum speed. The motor speed is then automatically varied between minimum speed and maximum speed in response to a proportional signal from the transducer. If the flow into the wet well increases beyond the capacity of the lead pump operating alone at maximum speed, the lead motor is transferred by a closed transition function from the inverter to the fixed-frequency a-c line to run at a constant speed. The lag 2 motor is then connected to the inverter and run at an adjustable speed as determined by the level in the wet well.

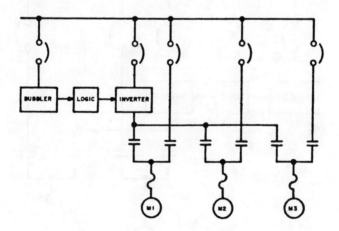

Fig. 17-27. Triplex One-Line

If the flow into the wet well increases beyond the capacity of the lead and lag 1 pump, the lag 1 pump is also transferred by a closed transition function from the inverter to the fixed-frequency a-c line to run at a constant speed. The lag 2 motor is then connected to the inverter and run at an adjustable speed as determined by the level in the wet well.

As the wet well decreases, the lag 2 motor shall decelerate to a preset minimum speed. As the level continues to decrease, the lag 2 motor shall be disconnected from the inverter and stopped. The lag 1 motor is then connected to the inverter by a closed transition function and run at a variable speed as determined by the level in the wet well.

The sequence of operation continues until only the lead motor is operated at variable speed.

Modes of Operation—
Duplex and Triplex Systems

The standard logic for a pumping system provides four modes of operation. The modes of operation are obtained by the positioning of two selector switches: Normal Emergency and Hand-Off-Auto.

Normal-Automatic—Sequencing and speed control of the pumps is automatic in response to the output of the pumps controller, level or pressure transducer.

Normal-Hand—Sequencing and speed control is the same as in the normal-automatic mode except that the automatic reference signal is replaced by the signal from a door-mounted potentiometer.

Emergency-Automatic—Sequencing of the pumps is automatic but at constant speed only.

Emergency—Hand—Start-Stop control of the pump at constant speed by use of door-mounted Hand-Off-Automatic Selector Switches.

Constant-Speed Sequencing

Inherent in the design of the duplex and triplex system logic are circuits that allow the motors to be sequenced at constant speed in response to the automatic reference signal. This mode of operation is referred to as the emergency-automatic mode and is actuated automatically in the event of an inverter fault or manually by selector switch positioning.

The constant-speed sequencing circuits provide a time delay between the starting of the motors. This precludes the possibility of simultaneous motor starts when, after a power outage, the automatic reference calls for all pumps to be in operation.

Motor Fault

Standard duplex and triplex sequencing logic provides that in the event of a motor fault, defined as a thermal overload or the selector switch in the off position. The motor's starter is opened, disconnecting the motor from the power supply. The next motor in sequence will start only when the reference reaches the level required to start that motor.

The motor starter thermal overload relay requires a manual reset to restore operation of the motor. However, unless otherwise specified, resetting of the thermal device will result in automatic restarting of the motor.

Lead Pump Alternation

This feature provides for alternation of the lead pump on a manual or automatic basis by means of a door-mounted selector switch. Positioning the switch to a designated pump number will place that pump in the lead position. The pump schedule remains fixed until the selector switch is repositioned. Positioning the switch to "timer" causes the lead pump position to be automatically alternated each 24 hours among all the pumps in the system.

Logic is provided that prevents either manual or automatic alternation of the pumps when more than one pump is running. This precludes pump alternation during periods of peak demand.

During pump alternation, the system shuts down, makes the necessary logic change, and restarts with the next pump in the schedule as the lead pump.

Typical pumping schedules are shown below:

Duplex	
Lead	Lag
1	2
2	1

Triplex		
	1st	2nd
Lead	Lag	Lag
1	2	3
2	3	1
3	1	2

TYPES OF ADJUSTABLE-FREQUENCY DRIVES*

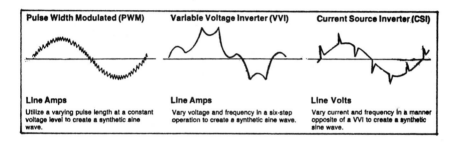

Fig. 17-18. Types of Adjustable-Frequency Drives

Each type of adjustable-frequency drive (AFD) has advantages and disadvantages. That is why Bell & Gossett (as well as other suppliers) offer a wide choice. The drive selection should be based on the needs of a particular ap-plication. To assist in your evaluation, eight basic AFD criteria should be considered.

Eight AFD** Basic Design Criteria

Efficiency—The supply current is changed from constant AC to variable DC and then inverted to a variable Hertz AC output. Each of these steps consume electrical energy dissipated in the form of heat. The less heat dissipated, the more efficient the AFD. Generally speak-

*This section is abstracted from Bell & Gossett Bulletin D 110A by permission.

**Note that in many cases, the term VFD (Variable Frequency Drive) is used.

ing, drive efficiency declines somewhat as speed is reduced. The importance of efficiency is often overstated. As the efficiency of the drive degrades with reduced speed, the system horsepower declines as the cube of speed, minimizing the effect of falling efficiency. This is shown graphically in the curve below.

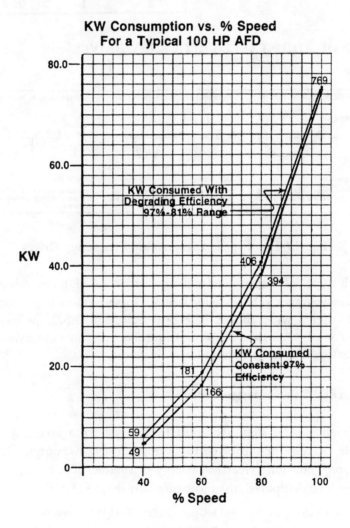

Fig. 17-19

Audible Noise—The switches that produce the synthetic sine wave may also produce an audible noise. The noise may be apparent in the drive itself or also in the motor windings with a PWM drive.

Size—Overall dimensions become an important factor in the mechanical room, where space is at a premium. If an AFD can be installed on a wall, valuable floor space is saved.

Cost—This is a major factor in AFD selection. Start with the first cost and add the cost of installation, drive maintenance, mechanical wear, downtime, and of course, the cost of operation.

Flexibility—This is a measure of the versatility of an AFD to fit into a variety of situations. Important features an AFD must handle include:

- Operation with standard NEMA B motors without specific motor tuning.
- No load operation for ease of setup and troubleshooting.
- Operation of motors with varying horsepower.
- Multiple motor operation.

Power Factor—This is the ratio between the actual power transmitted and the apparent power (voltage X current). Power factor is a function of the phase angle difference. A motor becomes unloaded as speed is reduced in a centrifugal application. The lighter the load, the lower the power factor. As in efficiency, the effect of a lowered power factor is minimized at reduced speeds due to the rapid decrease in horsepower.

Input Rectifier—This is the portion of the AFD that changes the 60 Hertz AC field supply into variable DC voltage. There are two common rectifiers—phase controlled and diode bridge. A phase controlled rectifier produces a degrading power factor with decreasing speed, but utilizes a less complex design with fewer components. A diode bridge rectifier maintains a uniformly higher power factor across the speed range. However, it requires a chopper circuit to provide a variable DC output which adds components and complexity to the circuit design.

Radio Frequency Interfence (RFI)/Electro-Magnetic Interference (EMI)—This describes the radiated and conducted interference generated by the switching circuitry in the rectifier. When sensitive computing or receiving devices are in the area, IEEE, FCC 15J Class

A, or VDE0871 specifications may be followed. If so, filters are available for use with Bell & Gossett AFD's which comply with these specifications. However, most applications will not be affected by RFI/EMI.

Bell & Gossett evaluated and tested many manufacturers and designs of AFD's and chose the PWM and VVI AFD as standard product offerings. These findings are illustrated below.

★★ Fair   ★★★ Good   ★★★★ Best

CRITERIA	PWM	VVI		CSI	
	Diode Bridge	Diode Bridge	Phase Controlled	Diode Bridge	Phase Controlled
Efficiency	★★★★	★★★	★★★	★★★	★★
Power Factor	★★★★	★★★★	★★★	★★★★	★★★
RFI/EMI Noise	★★★	★★★	★★★	★★★★	★★★
Audible Noise	★★	★★★	★★★	★★★★	★★★★
Size	★★★★	★★★	★★★	★★	★★
Cost	★★★★	★★★	★★★	★★	★★
Flexibility	★★★★	★★★★	★★★★	★★	★★

Fig. 17-20

*APPLICATION NOTES ON
VARIABLE FREQUENCY DRIVES

The Robicon Company has developed typical application data on AFDs. Obviously, all competitive brands may not have the same features, and investigation must be made in each case.

A. Torque Capability

The Robicon AFD can develop full-load torque from zero speed to 100 percent speed for a standard induction motor. A standard AC motor has thermal limitations which will only permit it to be operated *continuously* down to about 60 percent speed at full-load torque. If the motor is to be operated continuously at full-load torque below 60 percent speed, then the motor must be oversized or provided with external ventilation such as a blower. For a centrifugal load where the torque is reduced by the square of the speed, the AC

*Robicon Company information.

motor can be operated down to 10 percent speed continuously. The AFD also has the capability of providing an additional 25 percent starting torque boost for breakaway.

The AFD can run the AC motor at 10 percent overspeed or, as an option, 25 percent overspeed. The 10 or 25 percent overspeed range is a constant horsepower region. Since Hp = Torque X Speed, the torque capability is reduced in this overspeed range.

For applications requiring overspeed operation or low-speed constant torque operation, the horsepower required at the highest speed and the torque required at minimum speed must be specified. It is also important to specify if continuous operation is required at the minimum speed.

B. Multiple Motor Parallel Operation

Multiple motors can be connected in parallel to the AFD. The drive will run these motors at the same speed. There is a limit of 25 percent of the connected horsepower that can be disconnected while maintaining proper operation. For example, if four 100 hp motors are connected to a 400 hp AFD, 25 percent or one motor may be disconnected. If two motors are disconnected, the drive will not operate properly. Attempts to operate in this mode, however, will not damage the motors or the AFD.

If a parallel motor is disconnected while the AFD is operating, it is possible that an overvoltage trip may occur, and then the AFD will restart with the remaining motors. Motors being removed for maintenance should be disconnected from the AFD after it has been brought to a stop.

A motor can be reconnected to the drive only by turning the drive off, reconnecting the motor, and then restarting all motors together. If a motor is reconnected while the drive is running, the reconnected motor will draw very high inrush currents which may cause the drive to shut down.

C. Multiple Motor Sequential Operation

Sequential operation of multiple motors can be provided in a number of different system configurations. Contractors are usually involved. Closed transition transfer permits smooth transfer from the AFD to the AC supply and to the AFD from the AC supply. Automatic sequencing can be provided with the Robicon Logic-Mate Solid-State Sequencer.

D. Regeneration

The standard AFD is capable of regenerative braking up to 50 percent of the drive rating. The drive will automatically return power to the AC line to bring the motor to a stop in the time set on the deceleration ramp. Regeneration is available only for braking to a stop in the standard drive.

E. Variable-Frequency Drive vs.
   Slip Recovery Drive

Comparisons of variable-frequency drives (AFD) and wound rotor motor slip recovery drives (SRD) are made as alternative controls for variable-speed operation. The factors to consider when comparing the two drives:

1. AFD uses a standard induction motor which has a low initial cost, can be purchased with high efficiency characteristics, is readily repaired by a number of alternative sources, and can be purchased from many different manufacturers with relatively short lead time. Alternatively, the wound rotor motor is more expensive, less reliable, and can only be purchased from a small number of manufacturers.

2. A true comparison of the AFD and Slip Recovery Drive (SRD) must include the control, motor, input switchgear and installation costs. The SRD operates on the secondary windings of the wound rotor motor and requires a 3-wire connection from the secondary of the motor to the control. The wound rotor motor has primary windings that are connected

to the AC supply and this requires a combination starter and an additional 3-wire connection to the motor.

The AFD is connected between a standard induction motor and the AC drive and requires a 3-wire connection between the output of the control into the motor. An input circuit breaker is part of the standard AFD.

3. For a 480V installation, the AFD and motor will typically be less expensive than the SRD. For higher AC supply voltages, i.e., 2300 volts, the secondary of the wound rotor motor can be wound for 800 volts and the cost of the slip recovery control is reduced if the speed range is limited to 50 percent. The SRD then requires starting resistors and contactors to get to 50 percent speed. In this instance, the AFD and the SRD are competitive pricewise, but the SRD offers the disadvantage of the limited speed range, including no possibility of overspeed operation (10 or 25 percent overspeed).

4. Efficiencies of the AFD and SRD are comparable. The SRD manufacturers indicate that the power factor for their control increases as the speed is decreased. This statement is true but misleading. The power factor of the slip recovery control does increase from a very low value at full load and full speed. Ideally, one would like to have the best power factor under full-load conditions which is not the case with the SRD. The important consideration is not the power factor of the slip recovery control but the power factor of the system including both control and motor as seen by the AC line. Here the AFD power factor is better at all speeds than the SRD. In fact, power factor correction capacitors are typically recommended by the SRD because of the poor power factor at full speed and full load.

F. Variable-Frequency Drive vs. DC Drive

Variable-frequency drives for AC motors and DC drives for DC motors are alternative control choices for variable-speed operation. The following factors compare AC to DC.

The AC motor has these advantages over a DC motor:

1. More reliable—no brushes or commutator.
2. No scheduled maintenance required for brush adjustment or replacement.
3. Available from many manufacturers—generally has shorter deliveries.
4. Much lower initial cost.
5. Available in explosion-proof enclosures at higher horsepower rating.
6. Has smaller frame and lower WK^2 for equivalent horsepower rating.
7. Available at higher speeds in larger horsepower ratings.

The major disadvantage of the AC AFD as compared to the DC drive is that the AC control has more components than the DC control. The large number of solid-state components affect the AC control's reliability and cost largely offset by the DC motor itself. The relative reliability and expense of motor and control combination will depend on the environment.

The ability of the AFD to use a standard induction motor offers the normal advantage stated in "A" and in addition, the following:

1. Full-speed operation in event of an AFD failure. This bypass mode is not possible with DC motors.
2. Using the bypass ability, one AFD can be used sequentially to operate multiple AC motors. This reduces initial capital cost.

Existing constant-speed installations can be converted to variable-speed operation for energy-saving operation without changing existing AC motor.

For centrifugal loads, such as pumps and fans, DC motors may have excessive brush wear when operated at reduced speeds and light loads. This is not a problem with AC motors since they do not have brushes.

ISOLATION TRANSFORMERS

While it is not mandatory to use isolation transformers on AC equipment, there are certain conditions which require the use of isolation transformers. The use of isolation transformers offers the advantages of personnel safety, lightning protection, and equipment protection.

On ungrounded systems, lightning can induce high voltages on electrical equipment. If the electrical distribution system is a three-phase-wye configuration, the center of the wye is the best place to ground. This provides a path for the current from this induced voltage to go without flowing through the control. If this is not the case, and if the equipment is located in an area where lightning storms are a problem, a decision will have to be made. An isolation transformer would minimize the problem if the wye of the secondary were grounded. This again would provide a path to ground for the current caused by the induced voltage. However, if the wye of the isolation transformer is grounded, circuit common could not be grounded. Therefore, lightning protection could be achieved at the expense of personnel protection unless an isolation transformer was used in conjunction with a signal isolator. Under these circumstances circuit common and the wye of the transformer could be grounded and both lightning and personnel protection could be achieved.

In addition, it should be pointed out that even on grounded systems that include isolation transformers and signal isolators, there are still high voltage sections inside the control cabinet. Portions of the chassis, power bridge, and panel are high with respect to ground and only qualified personnel should maintain and troubleshoot the control.

On cascade systems, or on systems which must follow a grounded signal, isolation transformers or signal isolators are required on each of the drives. If the reference signals are not grounded, they should be isolated to withstand 600 AC RMS volts to ground.

Isolation transformers are to be used with all explosion-proof applications. A delta, wye transformer is suggested, 460/3/60 to 460/3/60 or 230/3/60 depending on primary voltage of controller.

Line pollution is an interesting phenomenon because no two installations have exactly the same equipment wired in exactly the same way. Drives normally do not excessively pollute the AC line and conversely, reasonable pollution on the line does not normally affect the drive. However, it is possible that the line may contain excessive pollution. This may necessitate the addition of an isolation transformer. The use of isolation transformers is the most effective means of eliminating the adverse effect of line pollution.

A ground fault can cause an A.F. control to fail just as it would a DC drive. This failure mode can be eliminated through the use of an isolation transformer in the controller input circuit. An isolation transformer is a must for all wash-down or "naturally wet" environments such as food processing, beverage processing, mines, etc.

PARALLEL PUMP SYSTEM CONSIDERATIONS

There are certain rules that need to be considered when applying VFDs for control of parallel pumps. These basic rules tend to govern parallel pump systems and are listed below, in order of decreasing importance.

Rule 1. A pump must not be operated continuously without flow through it.

Rule 2. The system should provide continuously variable control of total flow.

Rule 3. On-off cycling should be minimized.

Rule 4. Hydraulic and electrical transients should be avoided.

Rule 5. Frequent transfer should be avoided.

Rule 6. Long-term duty cycles should be balanced to equalize wear (alternation).

Rule 1 is fundamental to avoid over-heating the pump. It requires that a minimum speed adjustment be provided for VFD operation. The actual value will depend on the head across the pump, which in turn depends on the total flow. As the number of pumps in operation increases, the minimum speed value for the next pump must be increased.

Rule 2 implies that at least one of the operating pumps must be running from a VFD, although the minimum speed required by Rule 1 inevitably introduces a small discontinuity.

Rule 3 is of minor concern for VFD operation, but is very important for full-voltage starting of large motors. Where the system pumps out of a receiving tank in response to tank level, a discontinuity will produce a limit cycle whenever the input flow lies within the discontinuous range. The cycling frequency will be determined by the tank surface area, the flow difference across the discontinuity, and the hysteresis between on and off levels. When full voltage starting cannot be avoided, large amounts of hysteresis may be required.

Rule 4 also implies that full voltage starter control should be avoided. Sudden stops, in particular, can produce abrupt check-valve closures and severe water-hammer effects. The inrush currents associated with full-voltage starting are well known. Full-voltage control can be avoided even when there are fewer VFDs than motors, by means of closed synchronous transfer of a motor between the VFD and the power line.

Rule 5 does not contradict Rule 4 as much as it appears. Transfer will reduce system transients, but in exchange it introduces delays in the system response. These can also lead to limit cycling, especially in pressure-regulating applications. In general, therefore, use of transfer should be deferred until other means of avoiding transients are exhausted.

Rule 6 is usually satisfied by incorporating a means of periodically rotating the sequence in which the pumps respond to increases in demand.

18

Viscosity and Effect on Pump Design and Operation

There are some significant differences in pump behavior, and system response, when pumping fluids other than water. One of the major factors is *viscosity*. This is a characteristic of all fluids, including water, different for each fluid. It is of course obvious that fluids differ from solids; it is not so obvious that fluids can approach solids so that a distinction may be blurred. Fluids differ from solids in that a fluid continues to deform in the presence of a shearing stress. When a shear stress is applied to a solid, it deforms, then comes to a static equilibrium with the forces acting on it. In contrast, when a shearing stress causes a fluid to flow, the flow continues as long as the shear stress acts on it. By this definition, some substances, considered to be solids, are actually fluids. Tar, for instance, appears solid, but will flow under a shear stress. Even glass flows slowly, or creeps, when loaded with a shear stress. The resistance to shearing in the fluid is termed "viscosity." The term, at best, is only relative. It is a measure of the resistance to movement between adjacent particles in a liquid; more simply, it is a measure of the fluid's resistance to flow.

Viscosity should not be confused with adhesion, the ability of two unlike surfaces to stick together. Rather, viscosity is a measure of cohesion, the ability of two like surfaces to stick together. A general formula is:

$$\text{Viscosity} = n = \frac{F}{S} = \frac{\text{shear stress}}{\text{rate of shear}}$$

If we make the assumption that all materials at a given temperature have a viscosity that is independent of the rate of shear, we would be following in Isaac Newton's footsteps. Fluids with such characteristics are therefore called Newtonian fluids. There are, however, fluids which do not behave in this way, or in other words, their viscosity is dependent on the rate of shear. These fluids are called non-Newtonian fluids.

Three general categories of shear-related viscosity behavior are shown by Fig. 18-1. Note that the plots are straight lines because the shear rate is plotted on a log scale. These categories may simplistically be described as follows:

Newtonian: A fluid which has a constant viscosity at a given temperature regardless of the rate of shear. These liquids have a straight-line slope when plotting viscosity against temperature on the standard ASTM Viscosity chart. Water, most oils, and many common fluids are in this category.

Thixotropic: A thixotropic fluid is one in which the viscosity decreases as the rate of shear increases. There are relatively few liquids that are thixotropic enough to cause concern, but those that are, present peculiar problems. Typical of such fluids are many emulsions, rubber cements of some kinds, and ketchup. It is the thixotropic characteristic of ketchup that causes it to be difficult to start, but once started, it is difficult to stop.

Dilatent: A dilatent liquid is one in which viscosity increases with rate of shear. This characteristic is seldom a problem in pump selection, but when it is encountered it causes problems because the fluid becomes highly resistant to flow. Starches and clay slurries exhibit this characteristic.

The foregoing definitions are described as simplistic, inasmuch as there are subcategories in the non-Newtonian categories. Thixotropic

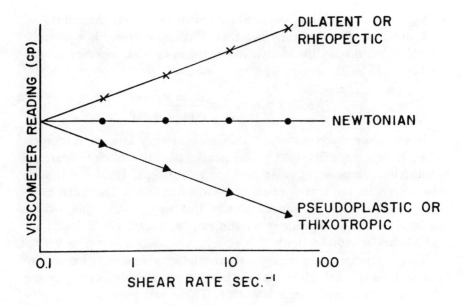

Fig. 18-1

and dilatent fluids are time independent. Rheopectic is the time dependent equivalent of the dilatent. Not only does the apparent viscosity increase with an increase in shear rate (or pipe velocity), but apparent viscosity also increases with time. The faster you shear it, and the longer you shear it, the more viscous it apparently becomes. This time dependency of such fluids presents a problem in long pipe lines, for instance, where the viscosity will change, depending on the distance of any fluid element from the pump. A time-dependent variation of a thixotropic fluid is known as pseudoplastic. In this case, the longer you pump it, the thinner it gets as illustrated in Fig. 18-2.

If a thixotropic fluid is subjected to a constant rate of shear for some time, its structure is gradually broken down and its apparent viscosity decreases to some minimum value. When the shear effect is removed and the fluid is at rest, the structure rebuilds gradually and

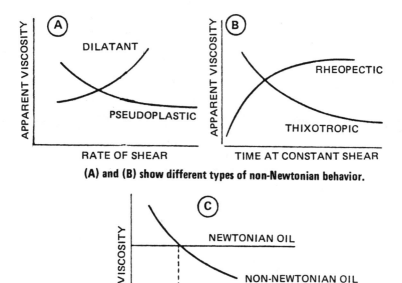

(A) and (B) show different types of non-Newtonian behavior.

(C) shows viscosity vs rate of shear for Newtonian and non-Newtonian oils.

Fig. 18-2

apparent viscosity increases with time to the original value. This is called *reversible thixotrophy*. If, however, upon removing the shear stress, a value less than the original viscosity is obtained with time, the phenomenon is known as *irreversible thixotrophy*. Some oils containing high-molecular-weight polymers and mineral oils at temperatures below their cloud point show this latter effect.

Quicksand is also thixotropic, since it becomes more and more fluid when agitated; therefore anyone caught in this water-and-sand mixture improves his chance of survival by remaining as motionless as possible.

If a rheopetic fluid is subjected to a constant rate of shear for a given period of time, its apparent viscosity increases to some maximum value. Upon cessation of shearing and resting for a time, its apparent viscosity decreases again.

As illustrated by Fig. 18-2, the determination of viscosity of a non-Newtonian liquid at only one shear rate is not usually sufficient. Incorrect conclusions would be drawn and application difficulties would be invited if the viscosities of a Newtonian and non-Newtonian oil were measured at some specific shear rate R_O, where the two curves happened to cross each other. While both oils have the same apparent viscosity at this one point, the remainders of their viscosity-shear curves are entirely different.

In addition to the above-mentioned phenomena of apparent viscosity, varying with shear and time, many fluids exhibit a yield stress below which no movement occurs. They exhibit rigidity as force is applied up to a certain degree, then break down, exhibiting usually pseudoplastic or thixotropic tendencies. They may be called Bingham plastics, false bodies, or whatever. It is obvious from this discussion that viscous fluids and slurries which exhibit viscous characteristics must be carefully studied before a pumping system is developed.

The two basic viscosity parameters are the Dynamic or Absolute Viscosity, μ, having the dimensions of force $\times$ time/(length)^2 and Kinematic Viscosity, v, having the dimensions of (length)^2/time. The parameters are related through the mass density of the fluid, p, such that $v=\mu/p=\mu g/y$, where y is the specific weight and g is the acceleration of gravity.

The unit of dynamic viscosity in English measurement is the pound-second per square foot (lb-sec/ft^2), which is numerically identical with the slug per foot-second.

The unit of dynamic viscosity in Metric measure is the dyne-second per square centimeter, called the Poise. This is numerically equal to the gram per centimeter-second. A convenient unit is the centipoise, which is 1/100 of a poise. Automotive industries use the unit called the Reyn, which is one pound-second per square inch. This unit is commonly used in lubrication problems.

The kinematic viscosity is the absolute viscosity divided by the density of the fluid. The unit of kinematic viscosity in English measure is the square foot per second. In Metric measure, the kinematic viscosity is expressed in square centimeter per second, called the Stoke. A more convenient unit is the centistoke, which is 1/100 of the stoke.

The two most frequently used viscosity units are the Saybolt Seconds Universal (SSU), and the centipoise (cp).

Viscosity in SSU is determined by the time (in seconds) required for a given quantity of liquid at a particular temperature to flow through a small short tube or orifice in a standard instrument. This timed flow measurement is the basic principle of the efflux-type viscometer.

Viscosity can be converted from centipoise to SSU by:

μ, in centistokes $= 0.226 \times SSU - 195/SSU$ for $SSU \leqslant 100$
μ, in centistokes $= 0.220 \times SSU - 130/SSU$ for $SSU > 100$
μ, in centistokes $= 2.24 \times SSF - 184/SSF$ for $25 \leqslant SSF \leqslant 40$
μ, in centistokes $= 2.16 \times SSF - 60/SSF$ for $SSF > 40$

Other conversions are shown by the table below:

Table 18-1

Multiply	by	to obtain
Poises	100	centipoises
pound-seconds/sq. ft.	47,880.1	centipoises
Reyns	6.89473×10^{-6}	centipoises
centipoises	2.08855×10^{-5}	pound-seconds/sq. ft.
centipoises	1.45038×10^{-7}	Reyns
Stokes	100	centistokes
sq. ft./second	92,903.4	centistokes
centistokes	1.07639×10^{-5}	sq. ft./second
sq. ft./second	$1488.16 \times \gamma$ in $\dfrac{\text{pounds}}{\text{cu. ft.}}$	centipoises
centipoises	$\dfrac{6.71970 \times 10^{-4}}{\gamma \text{ in pounds/cu. ft.}}$	sq. ft./second

Viscosities of all fluids are affected by temperature, for the viscosity of liquids becomes less as the temperature is increased. For this reason the temperature at which the viscosity is measured must be specified.

The Saybolt Viscosimeter, consisting of a calibrated orifice through which a flow of fluid is timed, provides arbitrary units of viscosity noted as Saybolt Seconds, and depending which orifice is used, the viscosity may be read as SSU, Universal, or SSF, Furol. Numerous other designations are also in current use, aimed for the convenience of designers working extensively with process fluids. A tabulation of common conversion terms is contained in Fig. 18-4, pages 324 and 325. Fig. 18-5, page 326, lists the viscosity and specific gravity of many common liquids.

A viscous liquid is usually defined in the pump industry as having a viscosity greater than 1000 SSU (approximately 200 cp). The Viking Pump Company considers a viscous fluid to have a viscosity above 700 SSU. The following paragraphs are from the Viking Pump Company literature:

Effect on Pump Installation—The viscosity of the liquid is a very important factor in the selection of a pump. It is the determining factor in frictional head, motor size required and speed reduction necessary. Frequently, for high viscosity liquids, it is more economical to use a large pump operating at a reduced speed since the original higher total installation cost is more than offset by reduced maintenance and subsequent longer life of the unit. Fig. 18-3 shows the percentage of rated used for pumping liquids of various viscosities.

Compared to other types of pumps, the rotary pump is best able to handle high viscosity liquids. The following tabulation shows the approximate maximum viscosity liquids that can be handled with various type pumps:

Centrifugal . . . . 3,000 SSU
Reciprocating . . . . 5,000 SSU
Rotary . . . . 2,000,000 SSU

The theoretical maximum allowable static suction lift is equal to 14.7 psi minus the frictional head. If the frictional head is high, an increase in suction piping size and port size will reduce the frictional

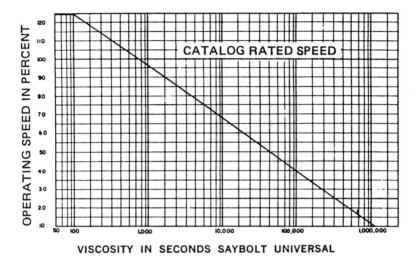

Fig. 18-3. Percentage of Rated Speed for Viscous Liquids
(Source: Viking Pump Company)

head and allow a greater static suction lift. On high viscosity liquids, the reduction of pump speed will also reduce frictional head and allow a greater static suction lift. Under some conditions, with high viscosity liquids, it may be better to relocate the pump to obtain a static suction head rather than to have a static suction lift. This relocation will help guarantee filling of the tooth spaces of the idler and rotor during the time they are exposed to the suction port and result in improved pump performance.

EFFECT OF VISCOSITY
ON PERFORMANCE

Liquids of high viscosity offer more resistance to flow than water. This increased friction occurs both in the piping and in the pump itself. More energy is required to force the liquid through the flow passages in the pump casing and through the pump impeller. The

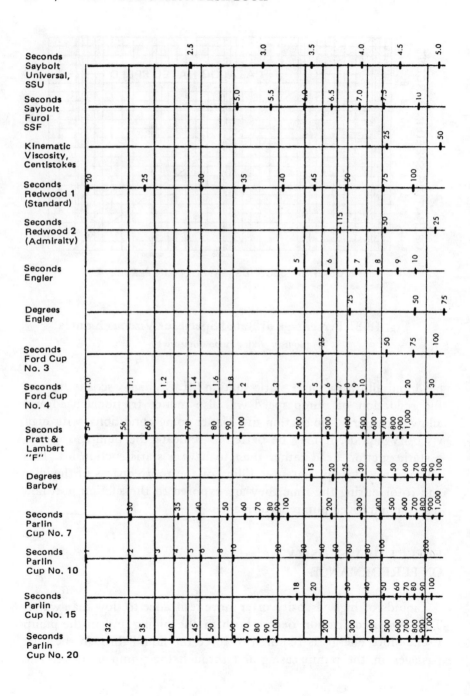

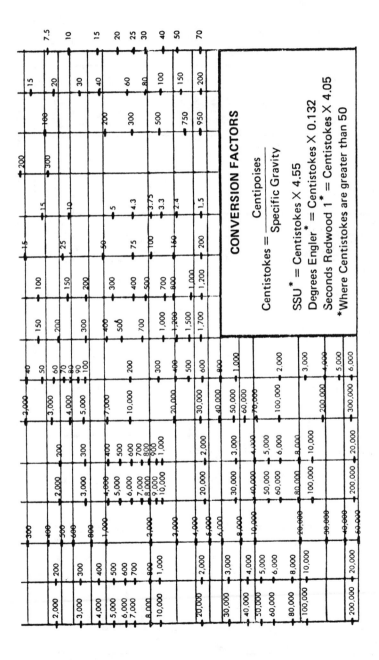

Fig. 18-4. Viscosity Conversion Chart

(Source: Viking Pump—Houdaille, Inc., A Subsidiary of Houdaille Industries, Inc.)

LIQUID	Spec. Grav.	Temp., °F	Viscos. SSU	Temp., °F
Asphalt Virgin*	1.03	60	7,500	250
			2,000	300
Blended RC-1, MC-1 or SC-1*	1.0	60	3,700	100
			1,100	122
RC-3, MC-3 or SC-3*	1.0	60	9,000	122
			3,700	140
RC-5, MC-5 or SC-5*	1.0	60	55,000	140
			4,500	180
Gasoline	.71	70	31	70
Glucose*	1.4	60	70,000	100
			7,500	150
Glycerine	1.25	70	3,000	70
			800	100
Glycol: Propylene	1.04	70	240	70
Triethylene	1.13	70	190	70
Diethylene	1.12	70	150	70
Ethylene	1.13	70	90	70
Milk	1.03	70	33	70
Molasses "A"*	1.43	60	12,000	100
			4,500	130
"B"*	1.45	60	33,000	100
			9,000	130
"C"* (Blackstrap)	1.48	60	130,000	100
			40,000	130
Oils Petroleum Crude (Penn.)*	.82	60	130	60
			60	100
Crude (Texas, Okla.)*	.85	60	400	60
			120	100
Crude (Wyo., Mont.)*	.87	60	650	60
			180	100
Crude (Calif.)*	.85	60	2,600	60
			380	100
No. 1 Fuel Oil*	.88	60	37	70
			34	100

LIQUID	Spec. Grav.	Temp., °F	Viscos. SSU	Temp., °F
No. 2 Fuel Oil*	.88	60	43	70
			37	100
No. 3 Fuel Oil*	.88	60	40	100
			36	130
No. 5A Fuel Oil*	.88	60	90	100
			60	130
No. 5B Fuel Oil*	.88	60	250	100
			175	130
No. 6 Fuel Oil*	.88	60	1,700	122
			500	160
SAE No. 10*	.91	60	200	100
			105	130
SAE No. 30*	.91	60	490	100
			220	130
SAE No. 50*	.91	60	1,300	100
			90	210
SAE No. 70*	.91	60	2,700	100
			140	210
SAE No. 90 (Trans.)*	.91	60	1,200	100
			400	130
SAE No. 140 (Trans.)*	.91	60	1,600	130
			160	210
SAE No. 250 (Trans.)*	.91	60	Over 2,300	130
			Over 200	210
Vegetable Castor	.97	60	1,300	100
			500	130
China Wood	.94	160	1,400	70
			600	100
Cocoanut	.93	60	140	100
			80	130
Corn	.92	60	140	130
			50	212
Cotton Seed	.90	60	170	100
			100	130
Linseed, Raw	.93	60	140	100
			90	130
Olive	.92	60	200	100
			110	130
Palm	.92	60	220	100
			125	130
Peanut	.92	60	200	100
			110	130

LIQUID	Spec. Grav.	Temp., °F	Viscos. SSU	Temp., °F
Rosin	.98	60	1,500	100
			600	130
Sesame	.92	60	190	100
			110	130
Soya Bean	.94	60	170	100
			100	130
Turpentine	.86	60	33	60
			32	100
Syrups Corn*	1.43	100	250,000	100
			30,000	130
Sugar	1.29 (60 Brix)	60	230	70
			90	100
	1.30 (62 Brix)	60	300	70
			110	100
	1.31 (64 Brix)	60	450	70
			150	100
	1.32 (66 Brix)	60	650	70
			200	100
	1.34 (68 Brix)	60	1,000	70
			280	100
	1.35 (70 Brix)	60	1,700	70
			400	100
	1.36 (72 Brix)	60	2,700	70
			650	100
	1.38 (74 Brix)	60	5,500	70
			1,150	100
	1.39 (76 Brix)	60	10,000	70
			2,000	100
Tar Coke Oven*	1.12	60	5,000	70
			1,000	100
Gas House*	1.24	60	150,000	70
			11,000	100
Road RT-2*	1.07	60	250	122
			60	212
RT-6*	1.09	60	1,500	122
			110	212
RT-10*	1.14	60	40,000	122
			300	212
Water	1.0	60	32	70

*Values given are average values and the actual viscosity may be greater or less than the value given.

Fig. 18-5. Approximate Viscosities and Specific Gravities of Common Liquids

disc friction of the impeller; that is, the resistance of the liquid to the rotation of the impeller, is greater when pumping a high viscosity liquid. These friction losses within the pump result in the pump generating a lower head and reduce the volume of liquid handled by the pump. These losses also result in a lowering of the pump efficiency and an increase in the brake horsepower requirements. The reduction in head, capacity, and efficiency and the increase in horsepower requirements of a pump handling a viscous liquid depend to a large extent on the type pump, the design of the passages inside the pump, the design of the impeller, the size of the pump, the capacity of the pump and the pump speed.

In recent years, much study has been given to the use of centrifugal pumps for viscous fluids. With the more sophisticated methods of analysis now available, viscous performance may be calculated from the performance with water. Pumps of the centrifugal type, radial, Francis, mixed flow, and axial flow are suitable for liquids of medium viscosities, and offer advantages over positive-displacement pumps. The differences in performance between pumping water and pumping a viscous fluid are illustrated in the curve of Fig. 18-6.

Fig. 18-7 provides a means of determining the performance of a conventional centrifugal pump handling a viscous liquid when its performance on water is known. It can also be used as an aid in selecting a pump for a given application. The values shown are averaged from tests of conventional single-stage pumps of 2-inch to 8-inch size, handling petroleum oils. The correction curves are, therefore, not exact for any particular pump.

When accurate information is essential, performance tests should be conducted with the particular viscous liquid to be handled.

Limitations on Use of Viscous Liquid
Performance Correction Chart

Reference is made to Fig. 18-7. This chart is to be used only within the scales shown. *Do not extrapolate.*

Use only for pumps of conventional hydraulic design, in the normal operating range, with open or closed impellers. *Do not* use for

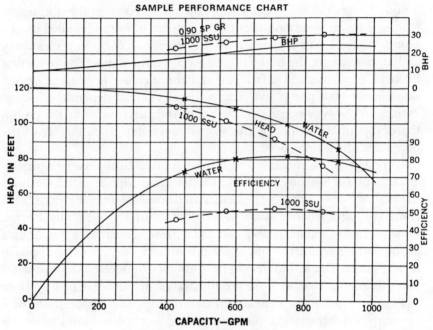

SAMPLE PERFORMANCE CHART

SAMPLE CALCULATIONS

	0.6 x Q_N w	0.8 x Q_N w	1.0 x Q_N w	1.2 x Q_N w
Water capacity (Q_w)	450	600	750	900
Water head in feet (H_w)	114	108	100	86
Water efficiency (E_w)	72.5	80	82	79.5
Viscosity of liquid.	1000 SSU	1000 SSU	1000 SSU	1000 SSU
C_Q—from chart.	0.95	0.95	0.95	0.95
C_H—from chart.	0.96	0.94	0.92	0.89
C_E—from chart.	0.635	0.635	0.635	0.635
Viscous capacity—Q_w x C_Q	427	570	712	855
Viscous head—H_w x C_H	109.5	101.5	92	76.5
Viscous efficiency—E_w x C_E	46.0	50.8	52.1	50.5
Specific gravity of liquid.	0.90	0.90	0.90	0.90
bhp viscous.	23.1	25.9	28.6	29.4

Fig. 18-6. Determination of Pump Performance
When Handling Viscous Liquids
(Source: *Hydraulic Institute Standards,* Twelfth Edition
By permission of Hydraulic Institute, 122 East 42nd St., New York, NY 10017)

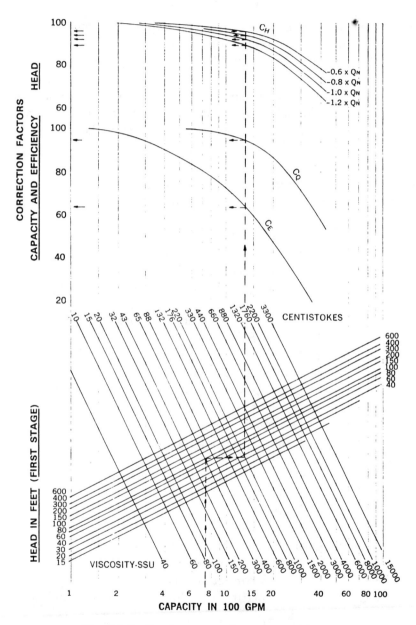

Fig. 18-7. Performance Correction Chart
(Source: Hydraulic Institute)

mixed flow° or axial flow pumps or for pumps of special hydraulic design for either viscous or nonuniform liquids.

Use only where adequate NPSH is available in order to avoid the effect of cavitation.

Use only on Newtonian (uniform) liquids. Gels, slurries, paper stock and other nonuniform liquids may produce widely varying results, depending on the particular characteristics of the liquids.

Symbols and Definitions Used in Determination of Pump Performance When Handling Viscous Liquids

These symbols and definitions are:

Q_{vis} = Viscous Capacity, gpm
       The capacity when pumping a viscous liquid.

H_{vis} = Viscous Head, feet
       The head when pumping a viscous liquid.

E_{vis} = Viscous Efficiency, percent
       The efficiency when pumping a viscous liquid.

bhp_{vis} = Viscous Brake Horsepower
       The horsepower required by the pump for the viscous conditions.

Q_w = Water Capacity, gpm
       The capacity when pumping water.

H_w = Water Head, feet
       The head when pumping water.

E_w = Water Efficiency, percent
       The efficiency when pumping water.

sp gr = Specific Gravity

C_Q = Capacity correction factor

C_H = Head correction factor

C_E = Efficiency correction factor

$1.0\, Q_w$ = Water Capacity at which maximum efficiency is obtained

**Instructions for Determining Viscous Performance
When the Water Performance is Known**

The following equations are used for determining the viscous performance when the water performance of the pump is known:

$$Q_{vis} = C_Q \times Q_w$$
$$H_{vis} = C_H \times H_w$$
$$E_{vis} = C_E \times E_w$$

$$bhp_{vis} = \frac{Q_{vis} \times H_{vis} \times sp\ gr}{3960 \times E_{vis}}$$

C_Q, C_H and C_E are determined from Fig. 18-6 and Fig. 18-7 which are based on water performance. Fig. 18-8 is to be used for small pumps having capacity at best efficiency point of less than 100 gpm (water performance).

The following equations are used for approximating the water performance when the desired viscous capacity and head are given and the values of C_Q and C_H must be estimated from Fig. 18-8 or using Q_{vis} and H_{vis}, as:

$$Q_w(approx.) = \frac{Q_{vis}}{C_Q}$$

$$H_w(approx.) = \frac{H_{vis}}{C_H}$$

**Instructions for Preliminary Selection of a Pump
for a Given Head-Capacity-Viscosity Condition**

Given the desired capacity and head of the viscous liquid to be pumped, and the viscosity and specific gravity at the pumping temperature, Figs. 18-7 or 18-8 can be used to find approximate equivalent capacity and head when pumping water.

Enter appropriate chart at the bottom with the desired viscous capacity, (Q_{vis}) and proceed upward to the desired viscous head

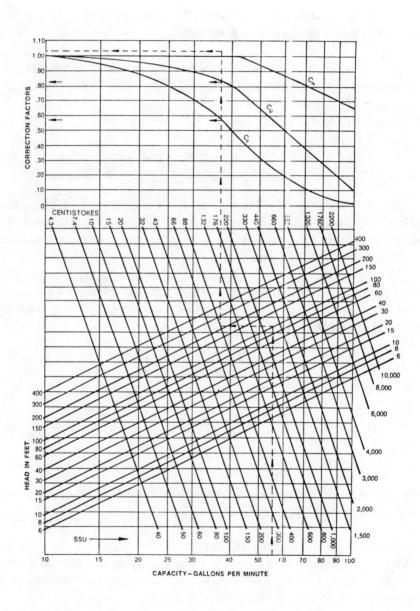

Fig. 18-8. Viscosity Corrections for Capacities of 100 gpm or Less

(H_{vis}) in feet of liquid. For multistage pumps use head per stage. Proceed horizontally (either left or right) to the fluid viscosity, and then go upward to the correction curves. Divide the viscous capacity (Q_{vis}) by the capacity correction factor (C_Q) to get the approximate equivalent water capacity (Q_w approximately). Divide the viscous head (H_{vis}) by the head correction factor (C_H) from the curve marked "$1.0 \times Q_w$" to get the approximate equivalent water head (H_w approximately). Using this new equivalent water head-capacity point, select a pump in the usual manner. The viscous efficiency and the viscous brake horsepower may then be calculated.

This procedure is approximate as the scales for capacity and head on the lower half of Fig. 18-7 or Fig. 18-8 are based on the water performance. However, the procedure has sufficient accuracy for most pump selection purposes. Where the corrections are appreciable, it is desirable to check the selection by the method described below.

Example: Select a pump to deliver 750 gpm at 100 feet total head of a liquid having a viscosity of 1000 SSU and a specific gravity of 0.90 at the pumping temperature.

Enter the chart (Fig. 18-7) with 750 gpm, go up to 100 feet head, over to 1000 SSU, and then up to the correction factors:

$$C_Q = 0.95$$
$$C_H = 0.92 \text{ (for 1.0 } Q_{nw})$$
$$C_E = 0.635$$

$$Q_w = \frac{750}{0.95} = 790 \text{ gpm}$$

$$H_w = \frac{100}{0.92} = 108.8 \cong 109 \text{ feet head}$$

Select a pump for a water capacity of 790 gpm at 109 feet head. The selection should be at or close to the maximum efficiency point for water performance. If the pump selected has an efficiency on water of 81 percent at 790 gpm, then the efficiency for the viscous liquid will be as folows:

$$E_{vis} = 0.635 \times 81\% = 51.5 \text{ percent}$$

The brake horsepower for pumping the viscous liquid will be:

$$bhp_{vis} = \frac{750 \times 100 \times 0.90}{3960 \times 0.515} = 33.1 \text{ hp}$$

For performance curves of the pump selected, correct the water performance as discussed below.

Instructions for Determining Pump Performance on a Viscous Liquid When Performance on Water is Known

Given the complete performance characteristics of a pump handling water, determine the performance when pumping a liquid for a specified viscosity.

From the efficiency curve, locate the water capacity $(1.0 \times Q_w)$ at which maximum efficiency is obtained.

From this capacity, determine the capacities $(0.6 \times Q_w)$, $(0.8 \times Q_w)$ and $(1.2 \times Q_w)$.

Enter the chart at the bottom with the capacity at best efficiency $(1.0 \times Q_w)$, go upward to the head developed (in one stage) (H_w) at this capacity, then horizontally (either left or right) to the desired viscosity, and then proceed upward to the various correction curves.

Read the values of (C_E) and (C_Q), and of (C_H) for all capacities.

Multiply each head by its corresponding head correction factor to obtain the corrected heads. Multiply each efficiency value by (C_E) to obtain the corrected efficiency values which apply at the corresponding corrected capacities.

Plot corrected head and corrected efficiency against corrected capacity. Draw smooth curves through these points. The head at shut-off can be taken as approximately the same as that for water.

Calculate the viscous brake horsepower (bhp_{vis}) from the formula given on page 331.

Plot these points and draw a smooth curve through them which should be similar to and approximately parallel to the brake horsepower (bhp) curve for water.

Example: Given the performance of a pump (Fig. 18-6) obtained by test on water, plot the performance of this pump when handling oil with a specific gravity of 0.90 and a viscosity of 1000 SSU at pumping temperature.

On the performance curve (Fig. 18-7) locate the best efficiency point which determines Q_{nw}. In this example it is 750 gpm. Tabulate capacity, head and efficiency for (0.6 × 750), (0.8 × 750) and (1.2 × 750). (See Sample Calculations, page 328.)

Using 750 gpm, 100 feet head and 1000 SSU, enter the chart and determine the correction factors. These are tabulated in the table of Sample Calculations. Multiply each value of head, capacity and efficiency by its correction factor to get the corrected values. Using the corrected values and the specific gravity, calculate brake horsepower. These calculations are shown on page 331. Calculated points are plotted in Fig. 18-8 and corrected performance is represented by dashed curves.

Fig. 18-8 is used in the same manner as Fig. 18-7 except that only one point on the corrected performance curve is obtained. Through the corrected head-capacity point, draw a curve similar in shape to the curve for water performance and having the same head at shut-off. If the capacity correction C_Q is less than 0.050, the corrected head-capacity curve should be a straight line. The corrected efficiency point represents the peak of the corrected efficiency curve, which is similar in shape to that for water. The corrected brake horsepower curves are generally parallel to that for water.

New Developments in Pumps
Handling Viscous Liquids

New developments are constantly modifying the restrictions which are supposed to ultimately limit the performance of pumps handling viscous material. In 1984, a wood products company was faced with the problem of handling hot viscous liquids, such as the molasses which hardboard manufacturers produce as a by-product of forming pressed wood-particle panels. The material is viscous, hot, and abrasive. The apparent viscosity of the liquid may be as high as 35,000 cps. At room temperature, the product, of 57 percent concentration, has the consistency of heavy ketchup. At elevated temperatures it flows like motor oil. Its non-Newtonian characteristics can cause erratic flow at the pump intake. A pump was supplied by the Worthington Company which solved the problems. At the pump end-suction intake, the nut holding the keyed-on impeller on the pump shaft has been removed and an inducer substituted. The inducer, an Archimedian screw projecting beyond the eye of the impeller, pulls the hot viscous liquid from the pipe intake into the suction port of the pump.

Positive-Displacement Pumps

For those who are planning to use positive-displacement pumps, the entire subject of selection, capacity determination, efficiency, and power requirements, are extensively covered in the "Viking Engineering Data" handbook, issued by Viking Pump-Houdaille, Inc., of Cedar Falls, Iowa 50613. The many uses of rotary positive-displacement pumps are not in this text, since the subjects are adequately discussed in the reference.

Viscosity Effects in Pipe Flow

At this point, it may be well to mention laminar and turbulent flow regimes. Turbulent flow is normally found when thin liquids move at fairly high velocities through a system. The flow pattern, as well as the distribution of velocity in the pipe, is entirely random, and unpredictable. Viscous fluids hardly ever attain turbulent flow.

Laminar flow is found when a thin liquid is moved at very low velocities through a pipe, or where a viscous fluid is moved. The flow particles within the pipe move in a straight line, but the velocity varies across the diameter of the pipe. The particles nearest the pipe wall move very slowly, while those in the center move rapidly. This flow pattern is predictable. The distinction between laminar and turbulent flow in the case of thin liquids depends on the Reynolds number of the moving fluid. This is a relationship between pipe roughness, fluid velocity, and viscosity, that predicts if flow will be turbulent or laminar.

Reynolds Number

In any fluid flowing in a pipe there is a continuous loss of head. This loss is the energy necessary to overcome the frictional resistance to flow. The loss depends on the size of the pipe, the velocity of flow, on the properties of the fluid, and usually on the roughness of the pipe. There are two distinct types of flow possible, streamlined flow and turbulent flow. The type of flow that will exist in a pipe is controlled by the velocity of the fluid, the shape of the pipe, the diameter of the pipe and the viscosity and density of the fluids. These variables have been combined into a ratio known as Reynolds number, Re:

$$Re = \frac{DVP}{N}$$

Where D is the diameter of the pipe or the hydraulic radius of the passage if it is not circular, V is the velocity, P is the density, and N the viscosity. The actual units used in the above are immaterial as long as they are consistent since Reynolds number is dimensionless. When Reynolds number is below 2100 the flow will always be streamlined flow, with a Reynolds number above 4000 the flow will be turbulent, in the transition range between 2100 and 4000 the flow may be either turbulent or streamlined depending on the conditions which exist.

Friction of Fluids in Pipes

The loss in head due to friction of a fluid flowing in a pipe may be determined from the Fanning equation

$$F = \frac{2fLV^2}{g\,D}$$

Where F is the friction loss in feet of the fluid flowing, f is a friction factor determined experimentally which varies with the Reynolds number, L is the length of the pipe in feet, V is the velocity of the liquid in feet per second, g is the acceleration of gravity (32.2 ft/sec/sec) and D is the inside diameter of the pipe in feet. This equation can be used for any fluid for either streamlined or turbulent flow and is applicable to channels other than circular if the hydraulic radius is substituted for the diameter of the pipe.

The friction factor, F, in Fanning's equation can be taken from the chart given in Fig. 18-9.

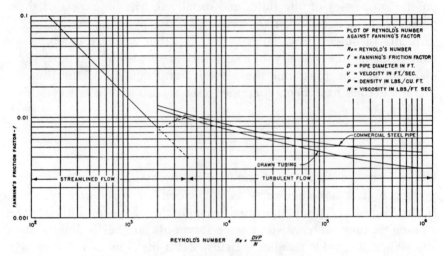

Fig. 18-9

(Source: Johnson Pump Co.)

It should be noted that current pipe flow engineering is no longer dependent entirely on the Fanning equation. Several other relationships have been developed which may better suit specific requirements. Nevertheless, the Reynolds number must still be used to determine if flow is laminar or turbulent. Virtually all pressure loss calculations for thin fluids are based on turbulent flow, while all calculations for pressure loss in a flowing viscous fluid are based on laminar flow.

19

Water Hammer

One of the persistent, and dangerous problems occurring when pumps are operated in extended pipelines, is the phenomenon referred to as "water hammer." The designation is apt. At the least, the piping will knock heavily, sometimes with considerable movement. At worst, the pipe will rupture violently. Several causative factors can be readily identified: a sudden stoppage of the flowing stream of water because of a quick valve closure; a reversal of the flow due to a pump shutdown; a separation of the flow stream downstream of a valve, due to the inertia of the fluid—the separation causes a vacuum which will quickly pull the stream backward, causing a high-velocity impact against the now closed valve; startup of a pump which injects a high-velocity stream into an empty pipe. It is obvious that the velocity of the stream, as well as its diameter, comprise the energy that, if interrupted, will be transferred to the nearest obstruction or containment. The problem is surmountable, but consideration must be given to piping configuration, operating conditions, and the time elements involved in valving, and pump start or shutdown.

CAUSATIVE ANALYSIS

The best analysis of water hammer considerations is made by the use of the Elastic Wave Theory. The accuracy of this wave analysis has been verified by many experiments, some of them dating back to the 1890s. The theory states that pressure changes in a closed conduit result from changes in flow which set up both positive and negative elastic waves. The summation of these waves establishes both the amplitude and shape of the resulting combined pressure wave. To compute this, however, certain fundamental factors must be realized:

• The pressure rise, when cutting off the flow instantaneously, is independent of the normal pressure and of the length of the conduit. Furthermore, for effective closures equal to the critical closure time of the conduit, or less, the pressure rise will be the same as for instantaneous closures.

• Surge pressures in closed conduits are directly proportional to the rate of change in flow velocity, and also to the velocity of the pressure wave transmitted along the conduit.

• The velocity of the pressure wave depends on the elasticity of the conduit and the compressibility of the liquid.

• For effective closures longer than the critical closure time, the pressure rise will be reduced by reflected waves, and the degree of reduction will depend on the interval of closure, and the nominal operating pressure within the conduit.

Proper application of these variables will explain the effects due to conduit material, branch piping, frictional losses, fluid characteristics, as well as other possible variables.

Constants Used in This Chapter

a = velocity of pressure wave in ft/sec
d = inside diameter of pipe in inches
e = thickness of wall in inches
g = acceleration of gravity, 32.2 ft/sec^2
E = Young's Modulus (see text)
H_r = pressure rise in feet above nominal pressure

k = compressibility of fluid (see text)
L = length of conduit in feet
P_1 = nominal pressure in conduit, psig
P_2 = surge pressure in conduit
P_r = pressure rise in psi above nominal pressure
V = velocity of fluid in ft/sec
w = density of fluid
Q = water flow, gpm
Z = volume of surge vessel, cu in.
C = a factor that is the ratio function of P_2/P_1 (see text)

Surge Pressure

Pressure rise in feet, above the nominal pressure, with sudden closure, is given by:

$$H_r = aV/g$$

In psi, the formula becomes:

$$P_r = 62.4 \, aV/144g$$

The calculation may be further simplified by substituting a good mean value of 4500 fps for a, and 32.2 fps for g, resulting in:

$$P_r = 60.56 \, V$$

When P_1 is the nominal system pressure, and P_2 is the maximum pressure due to the surge, then:

$$P_2 = P_1 + P_r = P_1 + (60.56V)$$

The above formulae do not take into consideration the compressibility of the fluid, or the effect of pipe expansion. If we assume that the expansion of the pipe is much less than the compressibility of the fluid, and there is no elasticity in the pipe, an approximate formula may be used:

$$P_r = V\sqrt{w \, k/g}/12$$

This equation gives conservative results, and it does make allowance for fluid compressibility. The value of P_r for water is 63 V, and for oil with a specific gravity of .8 and k of 250,000, it is 53 V.

Considering that the pressure rise due to sudden closure of a conduit is some 60 times the flow velocity, it becomes obvious that some type of control is mandatory to reduce ultimate pressure to manageable limits. Indeed, control is far cheaper than accommodation.

Surge Wave Velocity

The velocity of the surge wave is of primary interest, since the interval required for the wave to travel from the closure to the end of the pipe and back to the closure is a critical factor in the pressure effects due to the closure. The formula for surge wave velocity is:

$$a = 4660 / \sqrt{1 + kd/Ee}$$

Adequately close values for k, the compressibility of the fluid are:

300,000 psi for water
230,000 psi for crude oils
130,000 psi for gasoline.

The values for E, Young's Modulus, may be found in many engineering handbooks. Common values for E are:

30,000,000 psi for steel
12 to 15,000,000 psi for cast iron
3,400,000 psi for asbestos cement
variable for reinforced concrete.

An average value for water is 4000 fps. The intensity of the wave, as well as its velocity will be modified by the piping conditions such as friction, number of turns, distance for the wave to travel. Is is conceivable that given sufficient friction forces opposing the fluid movement, any surge would damp itself out. However, don't count on it!

Sudden and Gradual Closure

The critical time for conduit closure is defined as the interim at which closure is slow if exceeded, and instantaneous or sudden if not attained. The term "sudden" is preferred to "instantaneous" as from a realistic standpoint, nothing is instantaneous. The term "sudden" provides some logic for calculation. In water hammer calculations, this critical time is often used as a unit of time, with N defined as the number of such time units.

It has been demonstrated that water hammer pressures are a function of the rate of change of flow. This is quite different from the rate of change of area or the time of stem travel on the valve actuating mechanism.

To relate the type of valve movement to the actual rate of change of rate of flow, an approximation can be made. If the total time of valve travel T_T is plotted against the pipeline velocity, an effective time T_E can be obtained. This is the time in which full flow would be cut off if the change in flow was directly proportional to the valve stroke. In this case, where closure rate equals flow cutoff rate, the equivalent time T_E would be the same as the total time T_T, and the ratio would be $T_E/T_T=1$. In most valves this ratio varies, ranging from 0.40 to 0.60. An average of 0.50 can be assumed for approximate studies.

The critical time T_C for a conduit is

$$T_C = 2L/a \text{ seconds.}$$

Note that T_C is the interface between "sudden" and "gradual" closure. When gradual closure, where $T_E > 2L/a$ seconds, the pressure rise due to stopping the flow is reduced significantly. Fig. 19-1 shows this relationship for various effective times T_E and for various pipeline constants K.

The critical time can be increased by increasing the pipe friction (adding length or diameter); by using more elastic pipe material; inserting a rubberized section; but most commonly by attaching an accumulator close to the valve.

Surge Compressor Sizing

A surge compressor, or accumulator is a pressure vessel in which a gas acts to absorb the energy in the flowing fluid as it gives up velocity. The gas is compressed to a pressure that may be limited to any

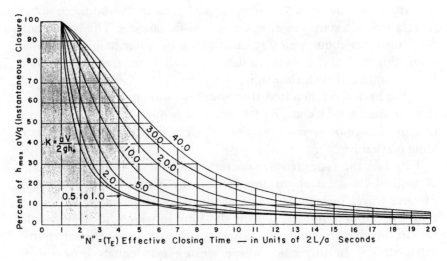

Fig. 19-1

desired value, depending on the volume of the vessel and the gas contained therein. The volume of the pressure vessel may be estimated by the expression:

$$Z = 62.4 \, A \, L \, V^2 \, C/P_1$$

The equation may be simplified and rewritten as:

$$Z = 0.057 \, L \, Q^2 \, C/P_1 \, d^2$$

Table 19-1 relates the factor C to the ratio of P_2/P_1. The volume of the vessel thus calculated will contain air or gas (if a bladder tank) that when compressed, will limit the system pressure to P_2 psi. It is axiomatic that the installation of surge relief devices is far cheaper than designing the entire piping system to take maximum pressures due to sudden fluid stoppage.

The previous discussion has not considered the effects of fluid compressibility or pipe expansion or movement. For precise calculations these refinements may be desirable. The effect of increased pressure is to cause the pipe to expand laterally, and because of

Table 19-1

C FACTOR AS A FUNCTION OF RATIO P_2/P_1	
P_2/P_1	C
1.25	1.03
1.50	0.6
1.75	0.42
2.00	0.34
2.5	0.23
3.0	0.2
4.0	0.16

Poissen's ratio, to shorten its length. If the pipe is straight between two well-anchored fittings, the tendency is to pull the pipe loose. If there is a bend in the pipe, the pressure on the bend will tend to lengthen the pipe, but the shortening will exceed the lengthening effect. If the angle is not anchored the pipe can bend and relieve the stress. When calculating these effects, the formulae for stress and strain are unnecessarily complex. The effect of pipe expansion being much smaller than that of compressibility of the fluid, a simple approximate formula may be used, assuming no elasticity in the pipe:

$$P_r = V\sqrt{wk/g}\,/12.$$

This equation gives values of P_r which are a few percent high, but which are therefore conservative.

PRACTICAL APPLICATIONS

The following paragraphs contain discussion of several common problems, which call for the installation of surge tanks and vents.

Many manufacturers have developed valves and valve control equipment devoted to specific aspects of water hammer control. One such system is a sophisticated relief valve that passes fluid into a precharged tank when line pressure exceeds safe values. Relief valves as offered for pressure limiting are also controlled electronically or under solenoid control. The use of solenoid control has the additional advantages that the surge valve may be given anticipating functions—that is, the valve can be opened or closed as required automatically when the pumps are started or shut down. Solenoid operators manipulate the control valves before surges have an opportunity to develop.

One other aspect of fluid surging which is not always addressed is the possibility of separation in the water column downstream of a closed valve, or as the result of pump shutdown. Water column separation, especially in large diameter piping, can develop sizeable vacuums, which may be sufficient to collapse the pipe. A common method of control is to provide a valve to bleed air into the line, thus minimizing the effect of the vacuum. Large, heavy-duty conduits should always be analyzed for possible collapse due to the development of significant vacuums.

The back and forth motion of the pressure wave is repeated a number of times, until friction damps it out. It is not unusual for the conduit to fail only after a number of repetitions of the surge have weakened the material. Surge tanks, or standpipes, require no outside attention, since they are self-contained, and are probably the most foolproof of all control devices. When air chambers are installed, it is very necessary for their proper operation to maintain the air in them at definite limits. Losses that occur because of absorption of air by water, or because of leakage, must be constantly replenished from an outside source. A water-logged air chamber is useless. Bladder-type chambers, in which air or gas is contained in a rubber bladder inside the tank, are finding favor, as the gas, being isolated from the water, is not lost.

WATER HAMMER DAMAGE IN
CENTRIFUGAL PUMP SYSTEMS*

Centrifugal pumps may be damaged by water hammer resulting from forces developed from inertia of pumped liquids when the pumps are stopped or started.

Newton's basic laws of mechanics state that a body (or matter) at rest will remain at rest, and a moving body will continue to move with constant velocity in the same direction, unless acted upon by an external force. The magnitude of force required to change the state of motion of a body is proportional to the mass and to the magnitude of the change in velocity during a given interval.

When a valve near the discharge end of a long pipe is closed, the mass of liquid flowing through the pipe will be stopped. This change in the state of motion can be accomplished only as a result of an applied force. That force may take the form of a sudden increase in localized pressure where the flow of liquid is blocked by a shutoff valve.

This pressure buildup will cause local compression of liquid as well as an elastic expansion of the housing surrounding the high-pressure zone. This elastic local deformation of the liquid and surrounding housing results in a pressure wave that travels through the entire pumping system (including the pump).

The magnitude of pressure buildup caused by closing a valve is proportional to the volume of liquid flow arrested, relative to the rate of valve closure. In long pipelines, a fast valve closure can generate sufficient pressure buildup to cause the pump casing to burst when the peak pressure wave rebounds.

The intensity of the pressure buildup is proportional to the valve closure rate; therefore, one effective method of preventing pressure buildup and resultant pump-casing failure is to close the valve slowly. This may be done manually or by using specially designed automatic controls. When neither method is desirable or practical, a relief valve can be installed between the valve and the pump.

---

*This section is from Worthington Pump Inc. *Pump World,* 1980 Vol. 6 No. 2.

A relief valve should be set to open at a pressure slightly above the system's working pressure. When the peak pressure wave reaches the relief valve, excess pressure (energy) dissipates. The valve returns to the closed position as pressure returns to normal.

Pump-casing rupture caused by water hammer can also be prevented by installing surge tanks and/or air vessels in the piping system. Surge tanks, properly sized and installed in the correct locations along the pipeline, will reduce the intensity of peak pressure waves and protect the pump.

Selecting the method to control water hammer requires consideration of:

• Cost of the installed system
• Cost and properties of the pumped liquid
• Size and layout of the pumping system
• Special requirements of the pumping system
• Physical arrangement of the pumping system
• Flow resistance through pumping-system components.

Pump damage resulting from inertia of the liquid is not limited to a buildup of excessive pressure. If pressure in any zone of the pump falls below the vapor pressure of the liquid, a cluster of vapor bubbles (or bubble) collapse vigorously, often causing serious mechanical damage. This action—called cavitation—may easily occur during a period of transient conditions when a pump is started or stopped.

The typical pumping system shown in Fig. 19-2 has a short suction line provided with a foot valve, a pump, and a long discharge line leading to a storage tank. In this system, the discharge valve is normally open; it is used only to prevent backflow when the pump is removed from the system for repair or maintenance.

After an operating pump has been shut down, inertia causes the rotating elements to continue to turn. The speed of the rotating parts decreases gradually until rotation stops. At the same time, liquid in the pipelines continues to flow as a result of inertia, and liquid continues to enter the suction line because of available NPSH (net positive suction head), replacing liquid discharged to the tank. This action continues for a short time after shutoff until the system comes to rest.

The head-capacity curve of a pump operating at a constant speed,

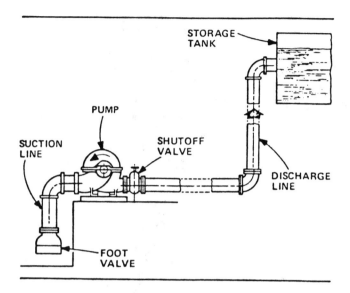

Fig. 19-2

A typical centrifugal-pump piping system includes a submerged foot valve on the suction line to prevent loss of prime, and shutoff valve in the discharge line to allow removing the pump for maintenance.

Fig. 19-3, is helpful in understanding this action. The most significant feature of this curve is that the head developed by the pump decreases as the flow rate increases. At a specific flow rate $Q=Q_0$, the pump ceases to develop head; at still higher flow rates, the head becomes negative. The pump, instead of adding energy to the flowing liquid, removes a portion of the total energy from the flowing liquid. Thus, total head of the liquid leaving the impeller is reduced to a value lower than the NPSH available at the suction intake.

When the pump speed is reduced, the relationship between head and capacity is also altered (shown schematically by the dashed curves in Fig. 19-3). Each reduction in speed results in a lower head-capacity curve. If a pump operating at a flow rate Q_a against a head H_a is shut down, then, after a time interval t, the pump speed will drop to $N=N_1$ and the head-capacity relation will be represented by the curve t.

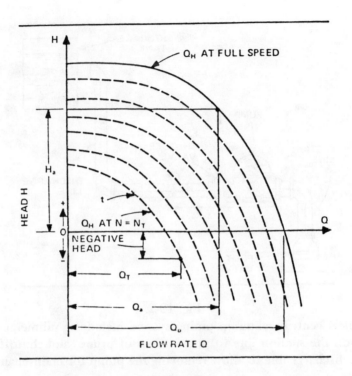

Fig. 19-3
Head developed by a pump operating at constant speed decreases as
the flow rate increases. If fluid inertia forces overrun the pump speed,
then energy is removed from the flowing liquid when the pump
operates as a motor.

At the same time, inertia of the liquid in the system causes flow,
although at a lower flow rate Q_t. Under certain circumstances,
the flow rate Q_t may still be sufficient to cause the pump to generate a negative head—that is, to dissipate energy at $N = N_t$. This discharge can result in a very low pressure at the discharge end of the
impeller, causing severe cavitation and/or erosion of the impeller
outlet, the pump casing, or both.

One of the simplest and least expensive ways to prevent cavitation
is to install an air valve on the discharge side of the pump. The air
valve is adjusted to admit air as soon as pressure at the impeller out-

let falls below a predetermined value. The vacuum is broken and cavitation is prevented.

In some installations, admitting air to the system is undesirable because of subsequent pump starting problems. In other instances, air is undesirable because it may change the properties of the pumped liquid. In such cases, other methods can be adopted:

- Increase the available NPSH by moving the pump to the lowest possible level or by adding a pressure reservoir.
- Install a flywheel on the rotating elements of the pump-driver combination, causing the pump speed to decrease more gradually.

A pump can also be seriously damaged during start-up. When a pump is started, liquid in the suction and discharge lines must be accelerated. The forces available to each mass, and those required for concurrent acceleration, may vary over a considerable range.

The force acting to accelerate liquid in the suction line is generated by the pressure differential between the total head existing at the suction inlet and the total head at the impeller inlet. The maximum value of this force, however, is limited to the difference between the NPSH available at the suction inlet and the NPSH required at the eye of the pump impeller. This establishes a low limit on the time required to accelerate liquid in the suction line. If liquid in the discharge line is accelerated at a rate faster than that defined by the minimum time required to bring the liquid in the suction line to full velocity, a vacuum will be created at the eye of the impeller, and cavitation may result. Cavitation at the eye of the impeller, resulting from inertia of the liquid in the suction line, depends on:

- Length and diameter variations of the suction line—time required for acceleration to required velocity depends on the total mass of liquid in the suction line.
- Difference between the NPSH available at the suction inlet and NPSH required at the eye of the impeller—pressure differential determines the rate of acceleration of the liquid in the suction line to full speed.
- Discharge line length—time required to accelerate liquid in the discharge line is proportional to the length of the line and is directly related to the time available for accelerating the liquid in the suction line without generating cavitation.

- Time interval required for rotating parts of the pump to accelerate to full speed—the shorter the time interval is, the less time is available for accelerating liquid in the suction line without cavitation. Flow resistance of particular sections of the pumping system may affect the length of the time interval during which transient conditions occur.

There are several ways to prevent cavitation at the impeller inlet caused by inertia of the liquid in the suction line:

- Select suction lines of large diameter and minimum length to reduce flow resistance.
- Maximize available NPSH by locating the pump as low as practical or by using a pressure reservoir.
- Open the discharge valve slowly to allow time for the liquid to accelerate without significant loss of available NPSH. (This procedure is not applicable to all pumps. Some pumps (for example, propeller pumps) develop instability and require increased power when operated against partially-closed valves. Because they must be operated against an open valve, another method must be used to prevent cavitation.)

Additional control methods include:

- Using a variable-speed-pump drive (automatic or manually controlled) that allows pump speed to increase gradually and the liquid in the suction pipe to accelerate to final velocity without a significant loss of available NPSH
- Installing an auxiliary tank at the pump inlet, and connecting it to the suction line with a pressure-actuated valve, Fig. 19-4

In the latter system, when pressure at the pump inlet falls below a certain limit, the valve opens and allows liquid from the tank to enter the low-pressure zone. With the tank properly located and proportioned, and the valve correctly set, cavitation can be prevented during startup.

Inertia during startup does not affect all pumps in the same manner. When operating at partial flow rates, propeller-type pumps are subject to high stresses caused by instabilities in the liquid flow. These stresses may result in severe damage. Propeller pumps are always started against an open valve to prevent damage. However, with long discharge lines, the inertia of the liquid increases flow

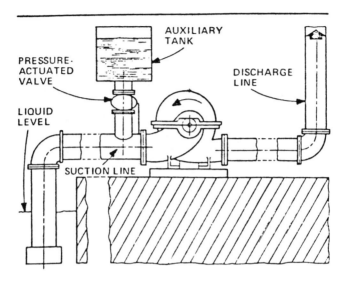

Fig. 19-4

Compensation for volume differential between flow in the suction and discharge lines may be made from an auxiliary tank, through a pressure-actuated valve near the pump intake port. Makeup liquid to the system acts similarly to a vacuum breaker, except liquid fills the void.

resistance during startup—the equivalent of starting the pump against a partially-closed valve. Damage to the pump can be prevented by installing a bypass piping system between the pump outlet and the check valve of the main discharge line, Fig. 19-5. The pump is started with the bypass valve fully open. The bypass system is closed gradually, redirecting liquid flow to the main discharge line while maintaining an adequate flow rate in the pump during the period of transient conditions.

In general, inertia is harmful only in pumping systems with long pipelines. However, in some systems with relatively short (30 to 45 ft) pipelines, inertia can cause serious pump damage through an effect often referred to as slam pressure. In a system consisting of a suction pipe (provided with a foot valve), a pump, and a discharge

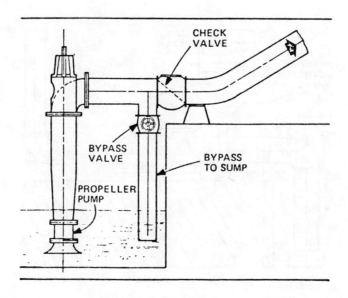

Fig. 19-5

Bypass piping system diverts a major portion of the pump discharge to the sump, allowing a gradual increase in flow through the discharge line. The bypass valve should be closed gradually until full flow from the propeller pump is transferred to the main line.

line delivering liquid to a tank, Fig. 19-6, inertia will cause the liquid in the system to continue to flow after the pump has stopped.

During this short period of extended flow, the foot valve remains open and liquid continues to enter the system, replacing the liquid that has entered the tank. When the liquid in the system comes to rest, the foot valve starts to close. In most cases, the disc or flap of a foot valve has poor hydrodynamic shape, and relative motion with the surrounding liquid creates high resistance. Therefore, some time after forward motion of the liquid has ceased, is required for the foot valve to close fully. During this period, the liquid starts to backflow through the open foot valve. This backflow eliminates relative motion between the disc (or flap) and the surrounding liquid, causing the valve to slam shut suddenly. The high speed of this

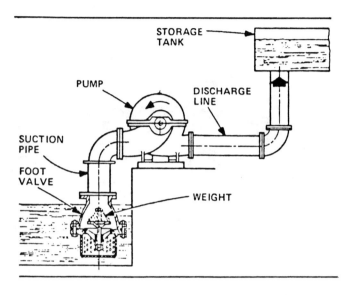

Fig. 19-6
Foot-valve operation can be modified to prevent severe closing action
on reverse flow. Preloading the valve with a weight or compression
spring detracts from the available NPSH.

closure creates an intense pressure wave, even in relatively short
pipelines, that may cause the pump casing to burst.

Slam pressure can be prevented by preloading the disc of the foot
valve with a weight, Fig. 19-6, or springs. Preloading causes the
valve to close before the returning flow of liquid has accelerated to
a significant velocity. This preloading is allowable only if reserve
NPSH is sufficient.

An alternate method is to install a relief valve in the suction line.
The relief valve is adjusted to open when pressure exceeds a predetermined value, allowing excess pressure to dissipate.

Slam pressure can also occur in the absence of a foot valve. When a
nonreturn valve is installed near the pump discharge, Fig. 19-7, inertia will cause the liquid between the nonreturn valve and the suction
line to continue to flow toward the suction tank, after the valve has

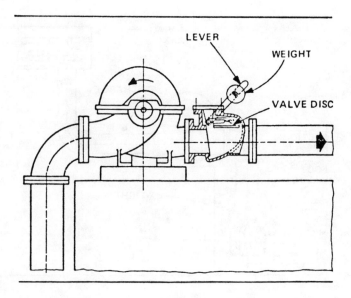

LEVER

WEIGHT

VALVE DISC

Fig. 19-7
Backflow and cavitation from inertia-induced vacuum in the pump can be prevented by installing a swing check valve close to the pump discharge. The action in this arrangement is parasitic to the discharge head.

slammed shut. This closure creates a vacuum in the pump and may lead to cavitation.

Preloading the disc of the nonreturn valve will cause it to close immediately after the liquid has stopped, thus preventing damage.

In a typical method, the valve disc is mounted on a long shaft extending outside the valve housing through a stuffing box. A lever is attached and loaded with a weight or spring, Fig. 19-7.

Another arrangement uses an automatic air valve, near the pump outlet, that will open as soon as the local pressure falls below atmospheric pressure, breaking the vacuum and exerting a cushioning effect.

WATER HAMMER CAUSED BY
AIR VENTING*

Operators of water pipelines comment that lines seem to be very susceptible to failure during the period they are being initially filled. The reason for some of these failures is apparent, but others are unexplained. Particularly difficult to explain are failures that occur out in the pipeline at high points where static pressures are minimum.

Some of the so-called unexplained line breaks appear to be water-hammer-type failures. Understanding how water hammer can occur during line filling can help identify these failures as being due to water hammer. Steps can then be taken to prevent the problem from occurring.

First, let's examine the basic water-hammer equation in simplified form:

$$H = 100 \ \Delta V$$

where   H = water hammer head, ft
ΔV = velocity change, ft/sec

From the equation, it is apparent that water hammer is created when there is a velocity change. A velocity change and, hence, water hammer can be originated by the venting of air from high points in the pipeline profile during line filling.

During line filling, air is vented to atmosphere through atuomatic or manual valves. The flow rate of fluid through an opening is proportional to the square root of the fluid density. The density ratio of water-to-air is approximately 820 to 1. The flow capacity through an opening is thus 29 times greater for air than for water. Due to this fact, when the water column reaches a vent valve during filling, an immediate decrease in column velocity is experienced—creating water hammer.

As an example, consider the pipeline shown in the sketch below. Assume the line is being filled at a velocity of 10 ft/sec as "there's

---

*This section and the following section are provided by Fluid Kinetics Corporation, Ventura, California.

no reason to fill slower for the line is open to atmosphere at the terminus and we can't have water hammer."

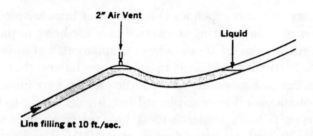

The venting capacity of the 2-inch air vent will permit air to vent matching the 10 ft/sec fill rate. At the moment the high point fills and the opening begins to flow water, the velocity change at this high point is essentially 10 ft/sec. This would cause a local water hammer of 1000 ft—certainly more than sufficient to cause a pipe failure.

A good engineering practice is to use small air vents so air is bled from openings over a sufficient time period so that joining liquid velocities are less than 0.5/ft sec. This will limit water hammer caused by air venting to 50 ft. A rule of thumb in sizing vent diameters is to make them less than .01 of the pipeline diameter.

CHECK VALVE SLAM

Some check valve advertisements promise that use of a certain check valve will control water hammer. This is correct only if we include the phenomenon of check valve slam under the general category of water hammer and exclude all other sources of water hammer. An understanding of the basic function and capabilities of check valves will explain why this is so.

Check valve slam occurs when the moving member of the check valve has sufficient energy to cause noise and movement of the valve body as it contacts the valve seat.

On pump shutdown, rotating inertia of the pump and driver causes the pump to maintain head and deliver fluid for a certain time period. When the head drops below the system head, flow reverses and the check valve closes.

Check valves have internal inertia and require a time period to move from open to close. If the check valve requires a greater time period to close than the time the pump inertia can maintain flow, the flow reverses before the valve is completely closed and the return flow catches the check valve moving element off the seat and slams it closed.

A typical system subject to check valve slam would be the water supply system for a high-rise building. Flow from the pump ceases very quickly due to the static head of the system, thus a valve that has the ability to close quickly is required. A valve properly used in such systems is the center-guided spring-loaded check valve as the valve disc is required to move only a short distance from open to closed.

For all systems, preference should be given to low inertia, quick closing pump check valves. These can be center-guided disc, tilting disc, or center-hinged dual-flapper type. Single-flapper cast-iron check valves have the poorest characteristics.

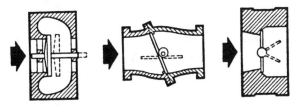

RECOMMENDED CHECK VALVES

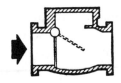

CHECK VALVE NOT RECOMMENDED

20

Super Speed Pumps
and Inducers

Within the last few years, pump equipment has become more versatile, with applications for specific classes of pumps which were not previously considered. In particular, suction conditions for centrifugal pumps have been analyzed to determine why they cavitate; additionally, centrifugals are being applied to difficult conditions, where NPSH may be a limiting factor. To extend the performance of superspeed pumps, an inducer, as shown in Figs. 20-1 and 20-2, is offered on a superspeed pump. The description is typical of other manufacturers. The curve of Fig. 20-3 shows the reduction in $NPSH_R$ due to the inducer.

Basically, the inducer is a very small booster pump, lifting the suction pressure to a sufficient level for the main conventional impeller. With this design, the booster pump and main pump are incorporated into one casing and mounted on a single shaft.

Fig. 20-4 shows the pressure distribution in an inducer and the associated conventional impeller. Liquid enters with suction pressure (or atmospheric pressure). Because of the small pressure drop at the inducer entrance, pressure never gets below vapor pressure.

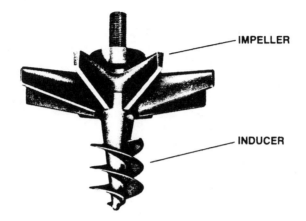

IMPELLER

INDUCER

Fig. 20-1

Fig. 20-2. High-Solidity Impeller

One benefit of the high-solidity impeller is a significant reduction in noise. A 10 dBA reduction in noise level will result in most instances, because the increased blade solidity also decreases the energy per blade, and fluid impinging on the stationary housings is less likely to set the pump case walls in motion. Less pump case wall vibration results in less noise being transmitted to the surrounding area.

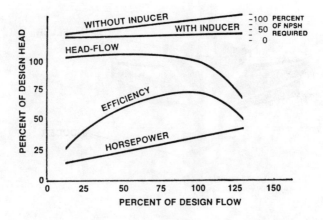

Fig. 20-3

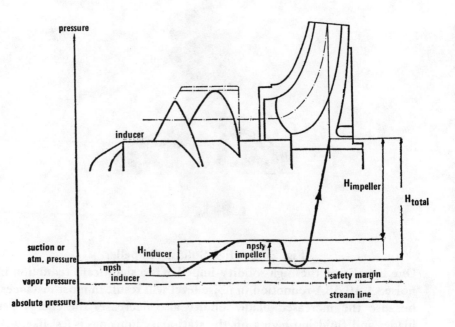

Fig. 20-4. Pressure Distribution in Inducer/Impeller Combination
(Source: Worthington Pump Inc., *Pump World*, Vol. 4, No. 3, 1978)

The inducer subsequently generates enough head to provide adequate suction pressure for the conventional impeller.

In spite of the large pressure drop at the conventional impeller inlet, the stream line never again approaches vapor pressure, avoiding any vaporization in the impeller. By increasing the generated head of the inducer, a large safety margin between the vapor pressure and the lowest pressure in the impeller can be achieved. The inducer/impeller combination has advantages of both.

The impeller and the inducer are designed with quite different assumptions. Each has certain advantages and disadvantages, but when the impeller and inducer are applied together, the disadvantages cancel out, and the advantages reinforce each other. Total head generation and efficiency are high, like a conventional impeller, but the combination also retains the low NPSH value.

As shown in Fig. 20-3, the inducer contributes some head to the total requirements, but only 5 percent or less. The efficiency of an inducer is, of course, lower than an impeller, but since such a small percentage of the total head is generated at this lower efficiency, overall efficiency of the pump doesn't change much. Overall bhp is not increased, and no power penalty has to be taken into account.

The inducer must be carefully integrated into the pump design, however. It is not sufficient to merely attach a screw to the impeller inlet. For instance, the inducer tends to show a region of rough operation at the low capacities. The onset of unstable operation can be controlled by the design of the inducer vanes, and the degree of pre-rotation to the conventional impeller. Worthington provides two inducer designs, therefore, for each impeller, to form a combination optimized according to the range of operation.

Before the application engineer can properly select an inducer pump, the minimum continuous capacity operation must be known. Different companies approach the problem in different ways. Fig. 20-5 shows the approach Ingersoll Rand uses, providing a bypass "stabilizer" to smooth the flow.

The Sundyne Company utilizes a "tuned" approach to inducer application. Suction specific speed is related to NPSH by flow and speed. However, in some instances, high inducer top velocities may

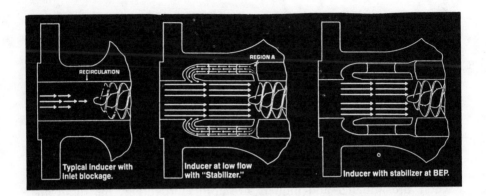

Fig. 20-5

(Source: Ingersoll Rand)

result in cavitation erosion. Many references indicate that velocity is the most significant parameter controlling cavitation erosion.

Usually, cavitation erosion is eliminated by decreasing pump speed until it is compatible with $NPSH_A$. This speed reduction may compromise pump efficiency because of a decrease in specific speed. Fig. 20-3 shows that in high-head, low-flow pumps, the maximum attainable efficiency increases with increasing specific speed.

The inducer method for eliminating cavitation erosion has been fully developed. Designing the inducer for the specific suction condition allows a reduction in the overall inducer diameter and therefore reduces tip velocity. Design considerations include adjustment of the inducer diameter and inlet blade angle relationship to minimize cavitation erosion. Traditional guidelines for radial equilibrium and diffusion factors limits are then used to complete the inducer design.

Inducer-equipped pumps, properly designed, will require less NPSH than similar pumps without the inducer.

Fig. 20-6 shows the performance envelope of a superspeed pump offered by the Sundstrand Corporation of Denver, Colorado. This chart illustrates the versatility of modern-day superspeed pumps; bear in mind that several companies now offer such units. For

PERFORMANCE CURVES REPRESENT RANGES OBTAINABLE WITH THREE
DIFFERENT IMPELLERS AND ONE CASE SIZE. CURVES ARE BASED ON
SPECIFIC GRAVITY OF 1.0.

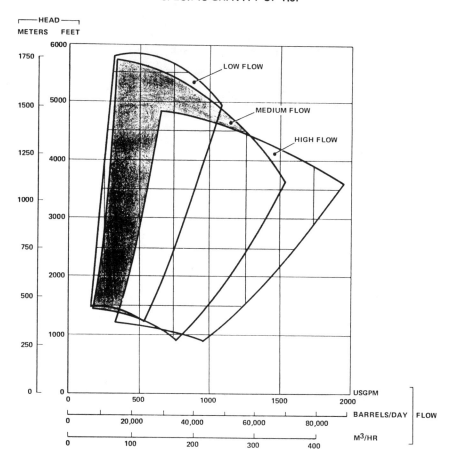

Fig. 20-6

(Source: Sundstrand Corporation)

instance, Fig. 20-7 is an illustration of a superspeed pump offered
by The Ingersoll Rand Company, having extremely high-head
capabilities.

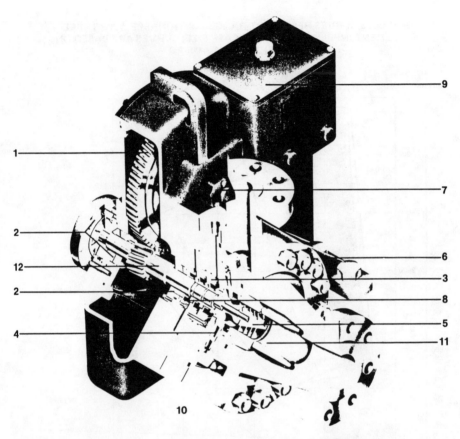

1 Single reduction helical gears deliver smooth, quiet operation. All gears are AGMA, class 11, carburized hardened and precision ground.

2 Rugged hydrodynamic radial and thrust bearings on high speed shaft for maximum dependability.

3 Specially designed mechanical seals in single, tandem or double arrangements. All API piping plans can be supplied.

4 Precision cast open or closed impellers sized for your specific application.

5 Various inducers available for low NPSH applications.

6 Centerline supported casing, stocked in carbon steel, 12 chrome steel and 316L stainless steel.

7 Primary oil pump for positive gear and bearing lubrication.

8 Separate oil seal integral with gear box.

9 Automatic prelube* for smooth, trouble-free starts.

10 Baseplates designed to API 610, 6th edition.

11 "The Stabilizer," a unique patented backflow catcher that insures stable inducer operation over entire head capacity curve.

12 Stiff shaft design insures operation at least 20% below the first critical speed.

*Patent Pending

Fig. 20-7

(Source: The Ingersoll Rand Company)

21

Standard Pumps Used as Hydraulic Turbines for Power Production

A power-producing hydraulic turbine, such as is used in utilities work, is similar in all respects, excepting installation and control details, to a pump in its dynamics, performance, and hydraulic application. It follows, therefore, that a standard production pump may be adapted to producing power wherever a head of fluid exists. Due to the large number of standard pumps produced, a turbine converted from a pump is cheaper than a specifically designed power turbine, by several orders of magnitude. A centrifugal pump may be run in reverse (both fluid direction and direction of rotation) with efficiency almost as good as it showed as a pump. The standard performance curves used for pumps may also be used to define operation as a turbine.

APPLICATION*

It is a reasonably simple matter to approximate the horsepower which can be recaptured from the potential energy available in a given situation. The key to the machine selection lies in the statement

---

*This section is abstracted from "Convert Pumps to Turbines and Recover HP," by Sheldon M. Childs, in *Hydrocarbon Processing & Petroleum REFINER,* Oct. 1962, Vol. 41, No. 10.

that *a good centrifugal pump operating with high efficiency may be expected to display good performance when the direction of flow is reversed and the pump is used as a driver.* Pump performance curves are readily available, so if we can evaluate our flows and pressure drops in terms of pump performance, selection of a pump for use as a turbine can be provided by any pump supplier.

Both pump and turbine efficiencies may be considered identical with little error, so the relation of performance between the two at their respective best efficiency points (BEP) and at the same speed becomes:

$$\text{Developed bhp Turbine} = \text{Required bhp pump or} \qquad (1)$$

$$\frac{Q_t \times H_t \times e \times Sp.Gr.}{3960} = \frac{Q_p \times H_p \times Sp.Gr.}{3960 \times e} \qquad (2)$$

Q = flow in gpm
H = Total head (pressure drop) in feet of liquid
e = The efficiency of the pump at its BEP

Subscript "p" refers to operation of a centrifugal pump; "t" refers to the same machine operated as a turbine.

Further

$$Q_t \times e = Q_p \qquad (3)$$
$$H_t \times e = H_p \qquad (4)$$

Recent investigations indicate that the above relationships are sufficiently accurate for actual pump selection from pump performance curves for applications as a hydraulic turbine.

Example 1. From a given process there is available 470 gpm fluid with 402 psig pressure drop. Sp. Gr. is 0.8. How much horsepower is available from a particular two-inch, four-stage centrifugal pump?

$$Q_t = 470$$

$$H_t = \frac{402 \times 2.31}{0.8} = 1160 \text{ ft of liquid}$$

e = 69 percent (from pump performance curves, Figs. 21-1 and 21-2)

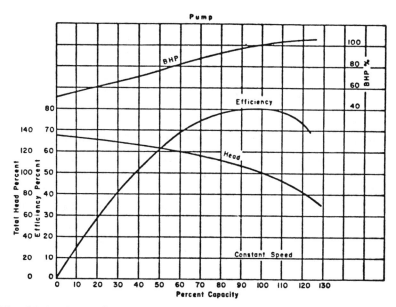

Fig. 21-1. Centrifugal Pump Performance Curves at Constant Speed

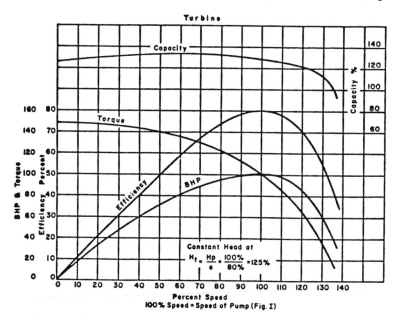

Fig. 21-2. Centrifugal Pump Performance Curves at Constant Head

From equation (2), $\dfrac{470 \times 1160 \times .69 \times .8}{3960} = 76$ bhp.

The pump selection was made at the following points:

$$Q_t \times e = 470 \times .69 = 325 \text{ gpm}, Q_p$$
$$H_t \times e = 1160 \times .69 = 800 \text{ ft}, H_p$$

At this point the actual pump selection has been made. A pump characteristic curve giving total head, efficiency and power versus capacity at constant speed should be obtained.

With the aid of Figs. 21-1 and 21-2, off-peak performance of the turbine can be estimated. The origin is a single-suction pump, with a specific speed of 1600 and a peak efficiency of 80 percent.

AFFINITY LAW

The normal application of a pump is at constant speed since motor drive is usually employed. Present applications of the hydraulic turbine indicate that a fixed pressure drop, or constant head is available. Therefore, speed has been used as the variable. This need not be restrictive since turbine operation is subject to the same affinity laws as apply to the pump.

$$\frac{Q_1}{Q_2} = \frac{rpm_1}{rpm_2}; \frac{H_1}{H_2} = \frac{rpm_1{}^2}{rpm_2{}^2}; \frac{bhp_1}{bhp_2} = \frac{rpm_1{}^3}{rpm_2{}^3}$$

Note: e remains constant

Example 2. This example indicates the application of the affinity law using values from Example 1. Additionally, assume values in Example 1 are at 3500 rpm and performance at 1750 rpm is desired.

Pump @ 3500 rpm — affinity step to — Pump @ 1750 rpm

$Q_p = 325$	e 69	$Q_p = 163$
$H_p = 800$		$H_p = 200$
$bhp_p = 76$		$bhp_p = 9.5$

Turbine @ 3500 — affinity step to — Turbine @ 1750

$Q_t = 470$	e 69	$Q_t = 235$
$H_t = 1160$		$H_t = 290$
$bhp_t = 76$		$bhp_p = 9.5$

as a check using equation (2),

$$\frac{235 \times 290 \times .69 \times .8}{3960} = 9.5 \text{ bhp developed}$$

Both speed points in Example 2 represent performance at the machine's BEP, however this stepping process can be made at any operating point, considering efficiency as remaining constant during the process. This permits adjusting the turbine performance curve to any desired speed range.

A more sophisticated method of analysis, taking into consideration the dimensionless characteristics of pump operation may be desired when performance must be determined in accurate detail, or when selection of a machine is in doubt, is as follows:[*]

PUMP AND TURBINE CHARACTERISTICS

Excepts from the complete characteristic curves of four different pumps are shown in Figs. 21-3 through 21-7 and 21-8 through 21-11. Figs. 21-3 through 21-7 show normal pump operation and Figs. 21-8 through 21-11 show normal turbine operation. The scales are dimensionless and chosen to facilitate the use of the affinity laws when making computations. The quantities n_n, Q_n, H_n, M_n, and P_n denote the speed, discharge, head, torque, and power of the pump at normal or rated conditions. The normal values used are those of the best efficiency point for normal pump operation. The quantities n . . . P without subscripts apply to any other condition of operation either as pump or turbine. The corresponding dimensionless quantities are defined by $v = n/n_n$, $q = Q/Q_n$, $h = H/H_n$, $m = M/M_n$, and $hp = P/P_n = mv$.

Figs. 21-3 through 21-7 show dimensionless head, torque, and power curves for normal pump operation plotted against q/v as argument. For constant-speed operation, v is constant and the curves

---

[*]This section is abstracted from "Centrifugal Pumps Used as Hydraulic Turbines," by C. P. Kittredge, ASME Paper No. 59-A-136.

Nomenclature

D = diameter, in.

H = pump or turbine head, ft of fluid

H_n = pump head at best efficiency point, ft of fluid

$h = H/H_n$ = dimensionless pump head

$hp = P/P_n$ = dimensionless pump or turbine power

M = torque of wheel on fluid, lb-ft

M_n = torque of pump impeller on fluid at best efficiency

$m = M/M_n$ = dimensionless torque of wheel on fluid

n = wheel speed, rpm

n_n = pump speed at best efficiency point, rpm

n_s = specific speed of pump (gpm basis) at best efficiency point

P = power to wheel, hp

P_n = power to pump impeller at best efficiency point, hp

Q = pump or turbine discharge, gpm

Q_n = pump discharge at best efficiency point, gpm

$q = Q/Q_n$ = dimensionless discharge

η_p = pump efficiency

η_T = turbine efficiency

$v = n/n_n$ = dimensionless wheel speed

represent the familiar pump characteristics normalized for the best efficiency point. The pump efficiency η_p is shown in decimal form. The DeLaval L 10/8 double-suction pump had the usual type of single-volute casing with side suction and side discharge. The Voith single-suction pump had a suction elbow of about 7.9 in. ID which was included in the measurements. The outlet diameter of the discharge nozzle was about 9.8 in. The overhung impeller discharged into a vane diffuser ring having a radial dimension of about $2^1/_8$ in. It had been built by the J. M. Voith Company as a model of a large unit designed for pump-storage service. Details of the Peerless 10 MH mixed-flow pump and the Peerless 10 PL axial-flow pump are shown in Fig. 21-6.

Figs. 21-8 through 21-11 show dimensionless head, torque, and power curves for normal turbine operation plotted against v/q as argument. Note that the values used to normalize these curves are

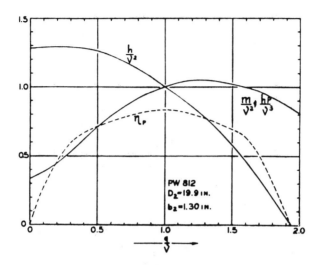

Fig. 21-3. Dimensionless Characteristic Curves of Normal Power Operation for DeLaval L 10/8 Pump. n_S = 1500 double suction.

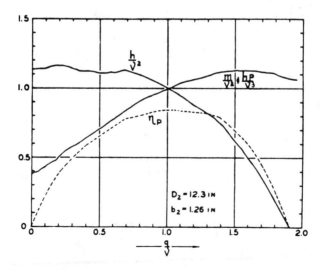

Fig. 21-4. Dimensionless Characteristic Curves of Normal Pump Operation for Voith Pump. n_S = 1935 single suction.

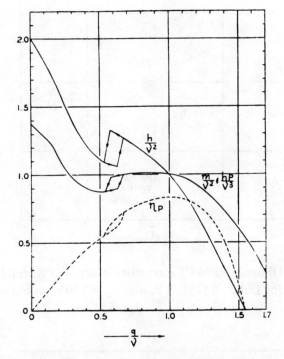

Fig. 21-5. Dimensionless Characteristic Curves of Normal Pump Operation for Peerless 10 MH Mixed-Flow Pump. $n_s = 7550$.

those of the best efficiency point for normal pump operation so that maximum turbine efficiency, η_T, does not coincide with v/q equal to unity. Figs. 21-8 through 21-11 are useful when selecting machines to meet specific requirements but, since turbines usually operate at nearly constant head, Figs. 21-12 to 21-15 have been included to show the performance curves of Figs. 21-8 through 21-11 plotted with v$\sqrt{h}$ as argument. The affinity laws used to convert the data of Figs. 21-8 through 21-11 to those of Figs. 21-12 to 21-15 are

$$\frac{v}{\sqrt{h}} = \frac{v/q}{\sqrt{h/q^2}} \tag{5}$$

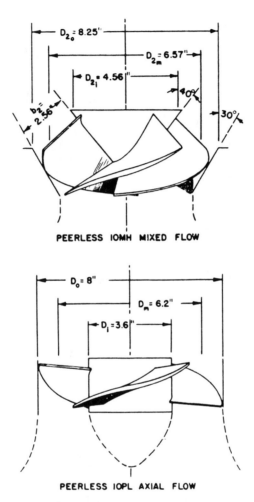

PEERLESS IOMH MIXED FLOW

PEERLESS IOPL AXIAL FLOW

Fig. 21-6. Dimensions of Peerless 10 MH and 10 PL Pumps

$$\frac{q}{\sqrt{h}} = \frac{1}{\sqrt{h/q^2}} \tag{6}$$

$$\frac{m}{h} = \frac{m/q^2}{h/q^2} \tag{7}$$

$$\frac{hp}{h^{2/3}} = \frac{hp/q^3}{(h/q^2)^{3/2}} = \left(\frac{m}{h}\right)\left(\frac{v}{\sqrt{h}}\right) \tag{8}$$

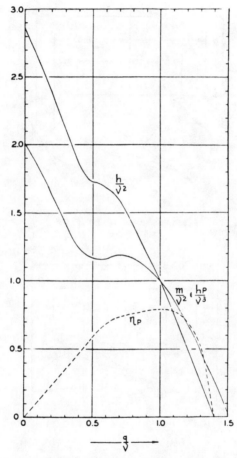

Fig. 21-7. Dimensionless Characteristic Curves of Normal Pump Operation for Peerless 10 PL Propeller Pump. n_s = 13,500.

SELECTION OF A MACHINE

A problem encountered in practice is that the head, speed, and power output of the turbine are specified and it is required to select a pump which, when used as a turbine, will satisfy these conditions. Assuming that performance curves for a series of pumps are available,

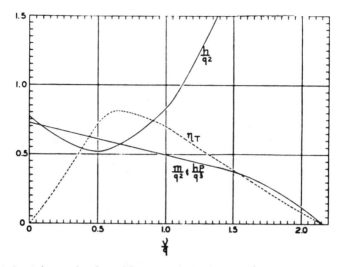

Fig. 21-8. Dimensionless Characteristic Curves for DeLaval L 10/8
Pump–Normal Turbine Operation

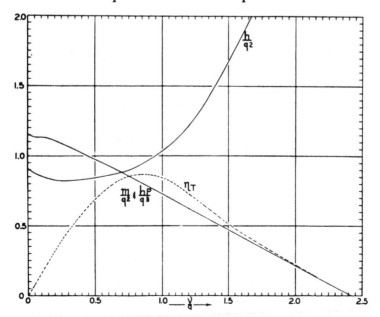

Fig. 21-9. Dimensionless Characteristic Curves for Voith Pump–
Normal Turbine Operation

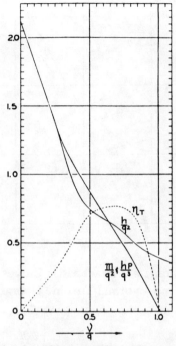

Fig. 21-10. Dimensionless Characterictic Curves for Peerless 10 MH Pump—Normal Turbine Operation

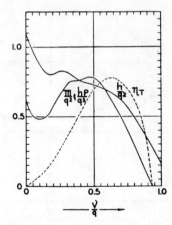

Fig. 21-11. Dimensionless Characteristic Curves for Peerless 10 PL Pump—Normal Turbine Operation

a typical set of such curves should be normalized using the head, power, and discharge of the best efficiency point as normal values. The normalized curves should be compared with those of Figs. 21-3 through 21-7 to determine which figure best represents the characteristics of the proposed pump. Once a choice has been made, the approximate turbine performance can be obtained from the corresponding figure of Figs. 21-8 to 21-11.

Assume, for example, that the turbine specifications are $H = 20$ ft, $P = 12.75$ hp, and $n = 580$ rpm; and that Figs. 21-3 and 21-8 are representative of a series of pumps from which a selection can be made. The turbine discharge in gpm is given by

$$Q = \frac{3958P}{H\eta_T} = \frac{(2958)(12.75)}{(20)(\eta_T)} = \frac{25,210}{\eta_T} \tag{9}$$

Only the normal values Q_n, H_n, P_n, etc., are common to the curves of both Figs. 21-3 and 21-8 so that these alone can be used in selecting the required pump. Values of v/q, h/q^2, and η_T are read from the curves of Fig. 21-8 and corresponding values of Q computed by Equation (9). Values of Q_n and H_n are then given by

$$Q_n = Q(v/q) \tag{10}$$

and

$$H_n = \frac{(H)(v/q)^2}{h/q^2} = \frac{(20)(v/q)^2}{. \, h/q^2} \tag{11}$$

For example, in Fig. 21-8 at $v/q = 0.700$ read $h/q^2 = 0.595$ and $\eta_T = 0.805$. By Equation (9), $Q = 25,210/0.805 = 3130$ gpm, by Equation (10), $Q_n = (3130)(0.700) = 2190$ gpm and, by Equation (11), $H_n = (20)(0.700)^2/(0.595) = 16.5$ ft. In similar manner, the locus of the best efficiency points for an infinite number of pumps, each having the same characteristics as shown in Figs. 21-3 and 21-8, is obtained and each pump would satisfy the turbine requirements. This locus of best efficiency points is plotted as curve A in Fig. 21-16. The head-capacity curve for a DeLaval L 16/14 pump having a 17-in-diameter impeller tested at 720 rpm is shown as curve B in Fig. 21-16. The best efficiency point was found to be at $Q_n = 3500$ gpm and $H_n = 37.2$ ft. The locus of the best efficiency points for this pump

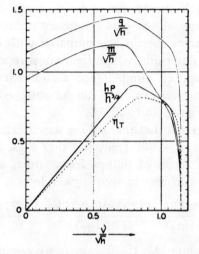

Fig. 21-12. Dimensionless Curves for DeLaval L 10/8 Pump—Constant-Head Turbine Operation

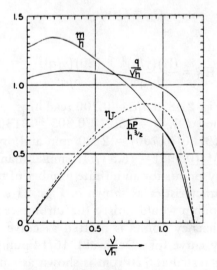

Fig. 21-13. Dimensionless Curves for Voith Pump—Constant-Head Turbine Operation

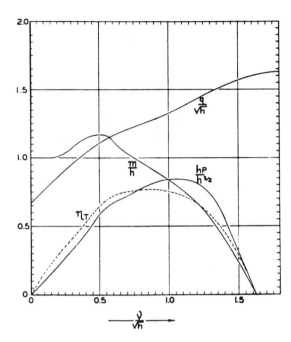

**Fig. 21-14. Dimensionless Curves for Peerless 10 MH Pump—
Constant-Head Turbine Operation**

for different speeds and impeller diameters is given by

$$H_n = (37.2)(Q_n/3500)^2 = (3.04/10^4)Q_n^2 \qquad (12)$$

and is shown by curve C in Fig. 21-16 Curve C intersects curve A at two points showing that the L 16/14 pump satisfies the turbine requirements. Only the intersection at the higher turbine efficiency is of interest. At this point $Q_n = 2400$ gpm, and $H_n = 18.8$ ft. Since the turbine speed was specified to be 580 rpm, the required impeller diameter is given by

$$D = \frac{(17)(2400/3500)}{580/720} = 15 \text{ in.}$$

or by

$$D = \frac{(17)\sqrt{18.8/37.2}}{580/720} = 15 \text{ in.}$$

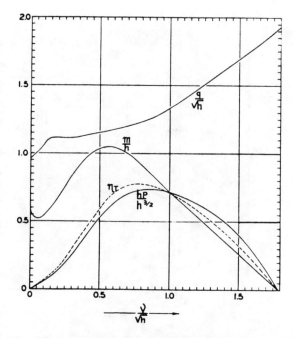

Fig. 21-15. Dimensionless Curves for Peerless 10 PL Pump—
Constant-Head Turbine Operation

The computed head-capacity curve for the 15-in.-diameter impeller at 580 rpm is shown as curve D in Fig. 21-16. The turbine discharge is 3200 gpm from Equation (9) with $\eta_T = 0.787$.

The optimum solution would be to have Curve C tangent to curve A at the point corresponding to maximum turbine efficiency, in this case $\eta_T = 0.812$. Since this would require a smaller pump, curve E in Fig. 21-16 shows a head-capacity curve for a K 14/12 pump, which was the next smaller pump in the series. Curve F, the locus of the best efficiency points, does not intersect curve A showing that the smaller pump will not satisfy the turbine requirements. It is important to note that the turbine head (20 ft) specified for this example was assumed to be the net head from inlet to outlet flange of the pump when installed and operated as a turbine.

The curves of Fig. 21-8 can be converted to show the constant-head characteristics of the pump when used as a turbine. The value

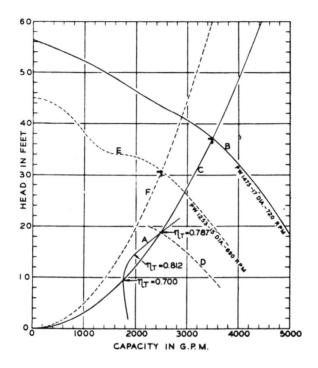

Fig. 21-16. Head-Capacity Curves for Example of Pump Selection

of $h = H/H_n = 20/18.8 = 1.064$ is constant for all values of the variables in Fig. 21-8. The speed is obtained from

$$v = n/n_n = (v/\sqrt{h})\sqrt{h}$$

and, solving for the speed

$$n = (v/\sqrt{h})(n_m\sqrt{h}) = (v/\sqrt{h})(580\sqrt{1.064}) = 698(v/\sqrt{h}) \qquad (13)$$

The discharge is obtained from

$$q = (Q/Q_n) = (q/\sqrt{h})\sqrt{h}$$

and, solving for the discharge

$$Q = (q/\sqrt{h})(Q_n\sqrt{h}) = (q/\sqrt{h})(2490\sqrt{1.064}) = 2570(q/\sqrt{h}) \qquad (14)$$

The power is obtained from

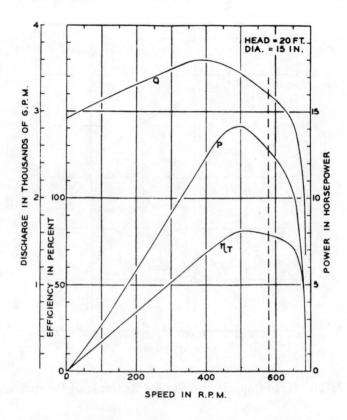

Fig. 21-17. Computed Constant-Head Turbine Characteristics for Delaval L 16/14 Pump

$$hp = (P/P_n) = (hp/h^{2/2})h^{2/2}$$

and, solving for the power

$$P = (hp/h^{2/2})(P_n h^{2/2}) = (hp/h^{2/2})(12.75)(1.064)^{2/2} = 13.39(hp/h^{2/2}) \quad (15)$$

The computation may be checked by

$$P = \frac{(Q)(H)(\eta_T)}{3958} \quad (16)$$

Fig. 21-17 shows the curves for the DeLaval L 16/14 pump with a 15-in.-diameter impeller as computed by Equations (13) through (16).

CONCLUSIONS

The performance of a given pump when used as a turbine can be estimated by the methods described. However, it probably will be necessary to apply the affinity laws over such wide ranges of the variables that the usual degree of accuracy should not be expected. Also, considerable care should be exercised if it becomes necessary to interpolate between Figs. 21-8 through 21-11.

It should be possible to select the correct size from a series of pumps to fulfill given turbine requirements even though the computed performance may differ from the results of subsequent tests.

Acknowledgments

The author of the paper wishes to thank Mr. W. M. Swanson who supplied the data for Figs. 21-5, -6, -10, -11, -14, and -15, and the DeLaval Steam Turbine Company, Trenton, NJ, for permission to publish Figs. 21-3, -8, -12, -16, and -17 and for assistance in the preparation of the figures.

The nomograph, Fig. 21-18 on the following page, is helpful in determining specific speed, and speed of rotation once the feet of head and the gallons per minute are determined. It will be useful in pump selection.

Specific Speeds of Turbomachines

Max Frey, Mechanical Engineer, Portland, Ore.

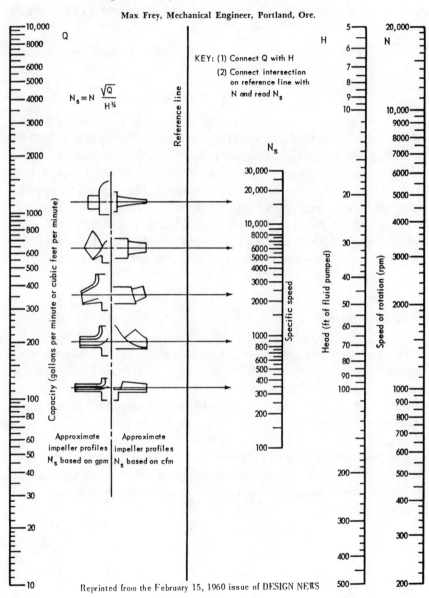

KEY: (1) Connect Q with H
(2) Connect intersection on reference line with N and read N_s

$$N_s = N \frac{\sqrt{Q}}{H^{3/4}}$$

Reference line

N_s

Capacity (gallons per minute or cubic feet per minute)

Approximate impeller profiles N_s based on gpm

Approximate impeller profiles N_s based on cfm

Specific speed

Head (ft of fluid pumped)

Speed of rotation (rpm)

Reprinted from the February 15, 1960 issue of DESIGN NEWS

Fig. 21-18

22

Pumping System Parameters

ACCELERATION HEAD REQUIREMENTS
FOR RECIPROCATING PUMPS

The flow variation resulting from the plunger action of reciprocating pumps requires that the fluid alternately accelerate and decelerate in the pump suction line. The energy required to accelerate the fluid in the suction line is usually referred to as "acceleration head."

Sufficient head must be available in the pump suction system not only to overcome suction system friction losses but also to provide the required acceleration head. If sufficient suction head is not available, the liquid will flash, forming vapor which is drawn into the pump cylinders. The shock pressures resulting from the sudden collapse of these gas or vapor pockets during the discharge stroke of the pump cause line vibration, noise and damage to the pump and connected components.

There is no universal agreement concerning acceleration head calculations. For relatively short, nonelastic suction lines, an empirical equation of somewhat general acceptance is available (reference Marks' Mechanical Engineers' Handbook, for example) as follows:

$$H_a = \frac{LVNC}{gK} \tag{1}$$

H_a = head in feet of liquid pumped to produce the required acceleration.

L = actual length of suction pipe in feet.

V = mean velocity of flow in the suction line in ft/sec.

N = pump rpm.

C = a factor for the type pump:

    C = .040 for quintuplex single-acting.

    C = .055 for sextuplex.

    C = .066 for triplex single- or double-acting.

    C = .115 for duplex double-acting.

    C = .200 for duplex single-acting.

    C = .200 for simplex double-acting.

g = 32.2 ft/sec^2.

K = a factor depending on the fluid:

    K = 1.4 for deaerated hot water.

    K = 1.5 for water, amine, glycol.

    K = 2.0 for most hydrocarbons.

    K = 2.5 for relatively compressible liquids such as hot oil, ethane.

In the foregoing, the following should be noted particularly:

1. V increases as pump speed is increased. H_a, therefore, varies with the square of pump speed.
2. H_a is greatest for liquids of relatively low compressibility such as water.
3. H_a varies directly with L.

The installation of a suitable stabilizer in the suction line near the pump reduces L to nil and thereby reduces the acceleration head requirement to nil. (Fig. 22-1)

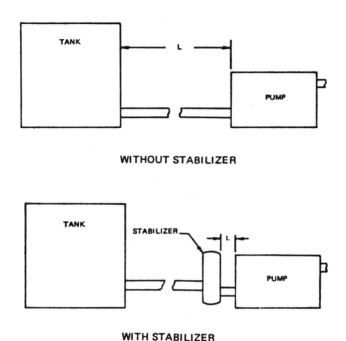

Fig. 22-1

(Source: Fluid Kinetics Corporation, Ventura, California)

EFFECT OF ELBOWS ON PUMP SUCTION

When liquid flows through an elbow, or from the run into the branch of a tee, both higher pressures and higher velocities are developed on the outside of the turn. This is not a contradiction of Bernoulli's Law, which applies to individual stream lines, because this phenomenon arises from a convergence of stream lines as a result of inertia. This results in fluctuating higher pressure and capacity on one side of the impeller at its inlet, and virtual starvation on the other side, which upsets the axial balance of the rotor and may cause cavitation on the low-pressure side. The result is axial oscillation of the rotor, overloading of the thrust bearing, noisy operation, and possible cavitation damage.

Installation of an elbow at the suction of a double-suction pump should be limited to arrangements in which the plane of the elbow is at right angles to the pump shaft. This applies whether the source of the liquid is above or below the pump and for bottom-suction as well as side-suction pumps.

In the case of single-suction pumps, whether side, top, or end suction, the orientation of an elbow at the pump suction is normally not critical, but may become so if the pump suction specific speed is over 10,000 in.

For most industrial centrifugal pump designs, suction flange velocities will vary between approximately 8 feet per second and 15 feet per second. The standard reducer between any two consecutive standard pipe sizes (between 10 inches and 30 inches) will reduce these values to ranges of approximately 4.5 to 5.5 feet per second and 8.4 to 10.4 feet per second. Below 10 inches, a one size reduction in diameter may affect a greater reduction of velocity; above 30 inches the effect will be less.

At a suction line velocity of 5 feet per second, a straight run of pipe equal to five pipe diameters should be adequate to rectify irregularities in the velocity profile which result from a 90-degree change in flow direction through an elbow or tee. But this may still be inadequate to stop a swirl generated by two or more such fittings in planes at right angles to each other. Under these circumstances, it may still be necessary to install straightening vanes at least two pipe diameters in length within the straight approach section.

At suction line velocities of 10 feet per second, the straight section will probably have to be at least 10 diameters in length, but above this value a straight approach should be provided as already discussed. Below this value, if an elbow must be used, it should be of the long-radius, constant-diameter type. Reduction from suction line size, if necessary, should be accomplished with a reducer downstream of the elbow.

In view of the problems inherent with elbows at pump intakes, Worthington engineers have developed a unique suction-elbow with geometry that will provide relatively straight flow and uniform velocity at the exit. The Worthington design (Fig. 22-2) uses a well-rounded orifice with the flow approaching from all sides. The area just preceding the outlet is larger than the outlet area. This increases the velocity from all sides, plus the acceleration, tends to straighten (or equalize) the flow in the elbow exit side. This design has been granted a U.S. Patent.

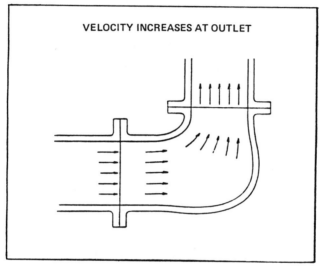

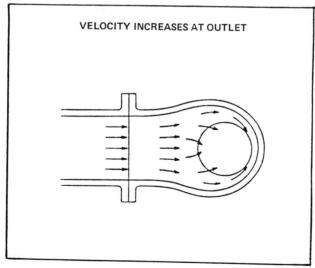

Fig. 22-2

(Source: Worthington Pump Inc.)

Other companies, for instance the Bell & Gossett Company, recommend a suction diffuser just upstream of the pump.

Reciprocating pumps require special consideration in piping to their suctions, because of the pulsing nature of the flow.

Many pump problems can be eliminated by the application of some simple guidelines during the piping system design stage. The guidelines listed below follow recommendations of the Hydraulic Institute and suggestions given by experienced piping designers.

Suction Piping

1. The piping should be as short and direct as possible. The number of turns should be held to a minimum and be accomplished with long-radius elbows or laterals.

2. The piping should be such that the natural frequency of any span is greater than 1.5 times the pump pulsation frequency.

3. The mean flow velocity should not exceed the following:
   2.0 ft/sec for pump speeds up to 250 rpm.
   1.5 ft/sec for pump speeds up to 330 rpm.
   1.0 ft/sec for pump speeds above 330 rpm.

   This will normally require suction piping one or two pipe sizes larger than the pump suction connection. An eccentric reducer should be used, with flat side up, to make the transition, as shown in Fig. 22-3.

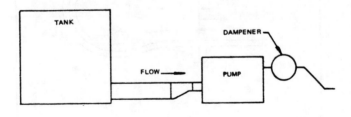

Fig. 22-3

4. The design should preclude the collection of vapor in the piping. If a high point is necessary, it must be vented.

5. The design should be such that the sum of acceleration head, friction loss, and pump NPSH required does not exceed the NPSH available from the system.

In equation form:

$$H_a + H_f + NPSH_{req'd} < NPSH_{avail.} \tag{2}$$

6. A suction stabilizer should be included adjacent to the pump suction, as shown in Fig. 22-4, if the requirements of paragraph 5 are not met. Make certain that the stabilizer is of suitable design for low pressure applications.

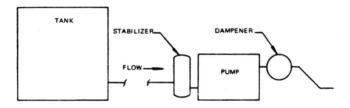

Fig. 22-4

Discharge Piping

1. Same as Item 1 for suction piping.

2. Same as Item 2 for suction piping.

3. The mean flow velocity should not exceed three times the suction line velocity.

4. A suitable pulsation dampener should be included adjacent to the pump discharge, as shown in Figs. 22-3 and 22-4.

VIBRATION OF PIPING IN
STEAM POWER PLANTS*

Vibration of steam power plant feedwater piping is always a source of annoyance and may be destructive unless the true cause of the vibration can be analyzed and eliminated.

Obviously it is best to eliminate these causes when the plant is being designed, but this becomes increasingly complex as power plants become larger, with attendant increases in length of piping, higher pressures from boiler feed and condensate pumps, and more feedwater heaters and tanks included in the entire system.

There are two kinds of vibration which may occur as a result of resonance—one, a function of mechanical characteristics of the piping system, and the other induced by internal hydraulic pulsations.

The first, or mechanically-induced vibration, will occur when a freely-hanging pipe has a natural frequency of vibration which corresponds to the speed of some rotating machinery in such a position that impulses may be transmitted to the affected pipe by interconnecting structural members.

Normally, this source of vibration is reasonably easy to diagnose although it may occasionally be complicated by the appearance of harmonic frequencies, which confuse the clear recognition of the source, and by the apparent absence of an interconnecting transmitter. At any rate, vibration in this category is easy to eliminate by "tuning it out." Since the vibrating element, or resonator, has a constant and determinable frequency (R), its natural frequency may be altered by stiffening, as in Fig. 22-5, or changing the length of the affected section.

Internal Pulsations

Second possible source of resonance, due to internal hydraulic-pressure pulsations, has a pump as the exciting element. The resonat-

---

*This section is abstracted from *Heat Engineering*, Nov-Dec 1959, "Does Cavitation Cause Feedwater System Vibration?" by A. Kovats. Reprinted from *Power Engineering*, Oct. 1959.

or is the entire piping system in which the elasticity of the water, the pipe and any vapor pockets in the piping, heaters, or other auxiliary equipment are the elastic elements.

It is generally assumed that centrifugal pumps have an absolutely steady flow, completely free of the pulsations which are well recognized in reciprocating pumps. Actually this is not entirely correct since the impeller blades moving past the diffuser vanes, or the tongue of the volute, always create a continuing series of pulsations of a constant magnitude.

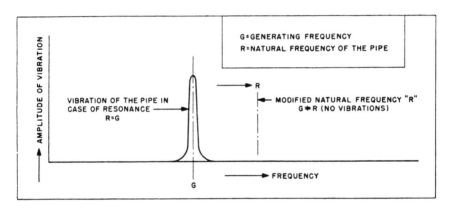

Fig. 22-5
Illustrating the simple case where vibration is caused from an external source such as rotating machinery. Cure here is easily accomplished.

These pulsations are of such a small intensity that they would not induce vibrations in a normal piping installation although it is possible that they might be amplified by any axial vibration of the rotating element, such as may occasionally be experienced in some boiler feed pumps equipped with balancing discs.

A much more important source of pulsation originates with cavitation—usually in condensate pumps. Cavitation is generally considered from the point of view of the expected damage to pump parts. This typical damage does occur when a pump is working at the point of incipient cavitation, close to the break point of the performance curve. When a pump is working in full cavitation, the vapor fills a considerable part of the space between the impeller blades and supplies a cushioning effect which reduces damage to a negligible amount. However, it is evident that the the flow and discharge pressure are no longer steady.

Some years ago it was accepted practice to operate condensate pumps with submergence control since this requires no auxiliary control equipment. When submergence control is used, the pump delivers the full capacity in accordance with its characteristic until the level in the hotwell of the condenser, and therefore the net positive suction head, drops below the limit that will prevent flashing in the suction eye of the first-stage impeller.

How It Works

When flashing starts, the flow of water is reduced which, in turn, reduces the system head and tends to increase the flow again. Therefore, the pump passes through the point of incipient cavitation and into full cavitation, as shown by the vertical breakdown line "A" of the characteristic performance curve in Fig. 22-6. The flow would then be established in equilibrium until there is some other disturbing influence.

If the amount of steam being condensed should be further reduced, there would be a further reduction in NPSH (net positive suction head) and another equilibrium operation established at some other flow, "B."

When a pump is operating in cavitation at some equilibrium condition, the flow may appear to be steady, but actually this can only be an apparent average. The collapse of the vapor pockets between the impeller blades is periodical, and as each pocket collapses, there is a corresponding reduction in occupied volume, resulting in an instantaneous reduction in velocity and pressure in the discharge piping. Therefore, the flow is never steady, and an apparently steady flow at capacity "A" in Fig. 21-6 is actually oscillating between the dotted lines as shown.

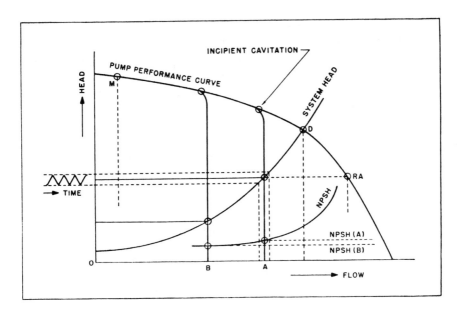

Fig. 22-6

Characteristic performance curves for a condensate pump illustrate
the relationships between the operation of the pump and the conditions of cavitation.

Variations in Frequency

Frequency of this pulsation is not constant, even for an equilibrium condition, because the dimensions of the vapor pockets do not
have precise geometric similarity. When there are two or more pumps
operating in parallel, each will have a slightly different mean frequency, and the combination may give an even more variable resultant which will be different for all combinations of flow and
temperature. It will be appreciated that this is a pulsation generator
without a measurable or calculable frequency. It can only be said
that it has a wide range which may be taken as "G" on Fig. 22-7.

Consider now this resonator composed of pipe, heaters, tanks,
water, and vapor pockets trapped within this system. Certainly, steel

has an important elasticity, and pipe bends will tend to act like the Bourdon tube in a pressure gauge. In most modern plants the elasticity of the water is a finite value, but the entrapped vapor pockets may have the most significant reaction. They act like springs, but the "spring constant" depends on the size, pressure and temperature of each pocket.

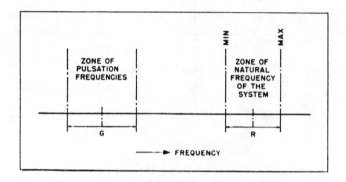

Fig. 22-7
Graph of Conditions Which are Necessary in Order to Prevent
Vibrations Caused by Hydraulic Pulsations

Frequency of pressure swings must depend on the water velocity, the length of the piping system, and the character of the water and vapor pockets in the system. The natural frequency may be calculated by using the formulae of Alievi and Bergeron. The longer the pipe the more extensive the vapor pockets for a given flow velocity and the lower will be the natural frequency. But in this case the resonator system also has a variable natural frequency, as indicated by "R" on Fig. 22-7.

Conditions Causing Trouble

When zones "G" and "R" are far apart there is no trouble. The typical frequency of pulsations of one condensate pump running in cavitation would be between 500 and 5000 cycles per minute. The natural frequency of a relatively short piping system including one or two heaters would be much higher—probably between 10,000 and 20,000 cycles per minute. This combination would not have much possibility of vibrating and is typical of the relatively old plants where submergence control was used for years and never gave any trouble.

Two pumps working in parallel would be much more likely to give trouble. If, as is the case in many new power plants, the pump, or pumps, may be of the vertical-pit type, the volume of the suction tank may appreciably reduce the natural frequency of the system and be more likely to result in vibration.

With a long and complex piping system, including several feed-water heaters, and particularly with a deaerating heater, the natural frequency, "R," may be so low that at some flow the zones "G" and "R" coincide. Under this condition, the usually insignificant pulsations may be multiplied in intensity to pressure peaks appreciably higher than the steady-flow discharge pressure. The pressure peaks are not liable to be dangerous in themselves, but the vibration which they create can be serious. (Fig. 22-8)

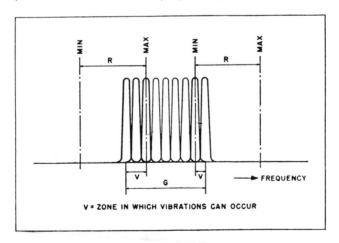

Fig. 22-8
This Graph Illustrates the Frequencies Where Vibration May Occur Due to Coincidence of Zones "G" and "R"

It is possible to make changes in the natural frequency of the piping system by adding surge tanks or altering the sizing of some of the pipe. This is rarely completely effective since the value of "R" may be changed only enough to be critical at some other flow. Also, additional anchors and stiffening of bends may stop the vibration, but this only suppresses the immediately apparent consequences of the pressure pulsations.

Obviously the only sure way of eliminating the tendency to vibrate is to eliminate the uncontrolled flow variations shown at "A" in Fig. 22-6. This means that submergence control must be discarded in favor of throttle or recirculation control.

In the case of throttle control, Fig. 22-9(a), an automatic float-operated valve, controlled by the level in the hot well, is used to adjust the flow. The pump is working along the performance curve between the points of design flow "D" and minimum flow "M" on Fig. 22-6. A small recirculating valve is also needed in this system to assure the required minimum flow. This valve must be opened when the flow is reduced to less than "M."

In the recirculation control system, Fig. 22-9(b), the automatic float-operated valve opens when the level in the hot well of the condenser drops and the excess capacity of the pump is discharged back into the condenser. If the steam consumption corresponds to flow "A" in Fig. 22-6, the pump will operate at point "RA" of the performance curve. The capacity difference "A-RA" is recirculated. Evidently at higher capacities than the design flow the pump needs more NPSH—otherwise it operates in cavitation.

Boiler Feed Pumps

In boiler-feed-pump service there should not be a similar problem if the NPSH has been correctly calculated, although an adequate margin above the minimum NPSH must be maintained to prevent cavitation during rapid load changes. Such temporary cavitation would also create pulsations which could induce serious vibration at certain critical flows.

It is also possible to trace the source of vibration to the boiler

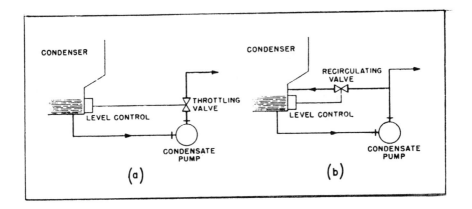

Fig. 22-9
Two Methods of Control Effective in Avoiding the Danger Zones
Shown in the Curves in Fig. 22-6

feed pump when the normal pulsations, amplified by axial vibration of the rotating element, happen to be in tune with the natural frequency of the piping system. This is more likely to be found when there is a deaerating heater in the system.

For example, in one case the axial vibration of the rotating part of a boiler feed pump could not be eliminated by any change of the balancing device. Since the vibration began at about half of the full capacity, increased with the flow, and coincided also with the vibration of the suction pipe line, it seemed possible that the source of vibration was outside the pump.

Frequency measurements indicated no relation to the speed of the pump. Opening the valve of a bypass line changed the frequency. Eventually, flared nozzles were installed in the outlet of the deaerator, as in Fig. 22-10, to damp the pulsations. This modification eliminated the vibrations.

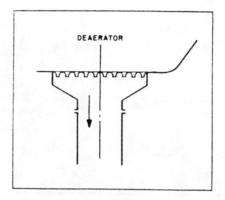

Fig. 22-10
Vibrations May Sometimes Be Decreased by Flared Nozzles
of This Type

Vertical Condensate Pumps

Sometimes it is difficult to distinguish between a mechanical and a hydraulic source of vibration without using proper instruments. Pump casings and motors of vertical condensate pumps are especially inclined to vibration. Unbalance of rotating parts and resonance with the speed of the motor can be easily detected since the frequency of the vibration corresponds to the speed of the motor and is independent of the flow.

But if the pump casing with the motor on top has a natural frequency only slightly different from the frequency of the motor or an outside source, as vibration of the base of the generator unit, a beat can result that may be mistaken for vibration resulting from pressure pulsations.

True Analysis Needed

The variation of the flow is linked with load variation and therefore with a slight variation of the speed of the motor. That is enough to change considerably the frequency of the beat and gives the impression that the vibration frequency varies with the flow. In such cases only by recording the vibrations and analyzing the curves can the true source of vibration be found.

In conclusion, it must be recognized that previous successful operation cannot always be accepted as the criterion for future design. Large modern power plants, with two or more condensate pumps in parallel operation in much more complex piping systems, are more susceptible to piping vibration.

Furthermore, even two presumably duplicate units will not have precisely the same characteristics. It is evident that it is always preferable to eliminate the cause and, therefore, the effect.

RESONANT VIBRATION IN
VERTICAL PUMPS

Variable-frequency drives applied to long-shafted vertical pumps have the same advantages as other process pumps; they also pose the possibility of resonant vibration. The Fairbanks Morse Company issues an application data sheet as follows:

Excessive vibration due to resonance may occur when a variable-speed drive is coupled to a pump. A resonant condition exists when the operating speed is at or near the natural frequency of the unit. When this happens, the inherent vibration of the unit is magnified and, as a result, destructive vibration may occur. This "unit" could consist of a variable-speed drive and high ring base (5710 construction), or could include a variable-speed drive directly coupled to the pump (5740 construction). (Fig. 22-11)

Types of variable-speed drives which can cause a vibration problem include magnetic couplings, hydraulic couplings, wound rotor motors, variable voltage and frequency controlled motors, vertical mounted offset belt drives, and vertical gear and horizontal engine combinations. Due to the wide operating range of a variable-speed driver, the probability of operating in a resonant condition is considerably greater than with a constant-speed drive.

To eliminate destructive vibration it is necessary to alter the

CONSTRUCTION SKETCH

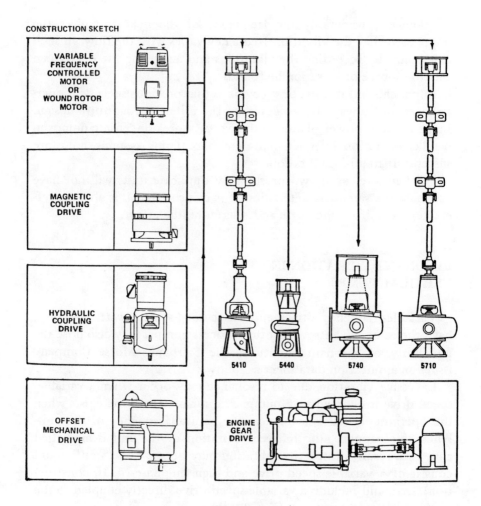

Fig. 22-11

natural frequency of the unit or limit the operating range. In most cases the latter is not practical. If the natural frequency is close to the upper end of the operating speed range (Fig. 22-12), the unit must be stiffened. This can usually be accomplished by providing lateral support at a point near the driver center of gravity.

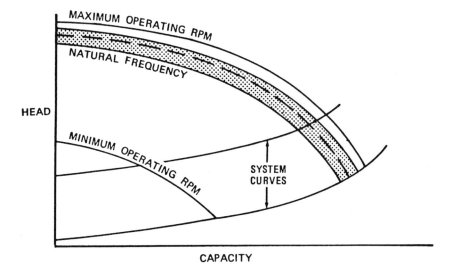

Fig. 22-12

If the natural frequency is close to the lower end of the operating range (Fig. 22-13), the unit may be made more flexible. This will lower the natural frequency below the operating range. During start-up and shut-down, the unit will pass through its natural frequency and vibration may occur. However due to the short duration of time this vibration should not be destructive.

The above "fixes" are very general and are offered as guides only. As these type of units are encountered they should be evaluated based on their own individual circumstances (operating speed range and equipment configuration).

When specifying these types of pumps for VFD, the specification should include a limitation of the critical reed frequency. This is the frequency of vibration of the column, typically about 600 cps maximum.

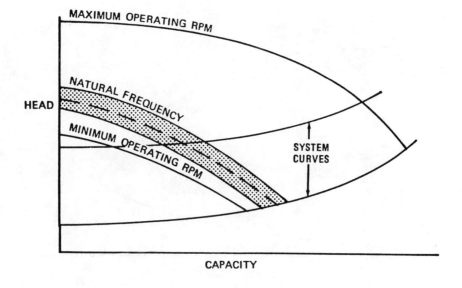

HEAD

CAPACITY

Fig. 22-13

EFFICIENCY AND POWER

Efficiency and power are indelibly linked together in pump design. The amount of power required to drive a pump depends not only on the hydraulic considerations, but on the loss of energy as the fluid goes through the pump. In a manner of speaking, the losses through the pump are not entirely wasted—any inefficiency shows up as a temperature increase between the suction and the discharge. Unfortunately, this temperature increase is usually of such low grade that it is not practically useful. It is a little known fact that the temperature at the bottom of Niagara Falls is greater than the temperature at the top. The difference may be measured, but may also be calculated, based on the amount of flow and the potential energy change as the water falls.

Efficiency of a pump is determined by many factors, including impeller shape, specific head, speed, volute or diffused design, and even the finish of the wetted elements of the pump. Wear will noticeably alter the efficiency, since clearances are changed. The brake horsepower to a pump is greater than the hydraulic horsepower due to these mechanical and hydraulic losses. Therefore the efficiency is the ratio of these two values:

$$\text{Pump Efficiency, e} = \frac{\text{whp}}{\text{bhp}} = \frac{Q \times \text{TDH} \times \text{Sp. Gr.}}{3960 \times \text{bhp}} \qquad (3)$$

The work performed by the pump is a function of the total head and weight of the liquid pumped in a given time period. The pump capacity in gpm and the liquid specific gravity are normally used in pump formulae rather than the actual weight of the liquid pumped. Pump input or brake horsepower (bhp) is the actual horsepower delivered to the pump shaft. Pump output or hydraulic horsepower (whp) is the liquid horsepower delivered by the pump. These two relationships are defined as follows:

$$\text{whp} = \frac{Q \times \text{TDH} \times \text{Sp. Gr.}}{3960} \qquad (4)$$

$$\text{bhp} = \frac{Q \times \text{TDH} \times \text{Sp. Gr.}}{3960 \times \text{pump eff.}} \qquad (5)$$

The constant 3960 is obtained by dividing the foot pounds equivalent to one horsepower (33,000) by the weight of one gallon of water (8.33 pounds). An alternate formula for horsepower is possible, considering that 2.31 feet of water at a specific gravity of 1 equals 1 psi. The formula then becomes:

$$\text{bhp} = \frac{Q \times \text{psi}}{1714 \times \text{pump eff.}} \qquad (6)$$

The formula using psi, while convenient, does not provide exact answers if the specific gravity of the fluid is not one. This error is noticeable in the case of high-pressure boiler feed pumps, which move hot water at specific gravities much less than one. The bhp values obtained by formula indicate the horsepower requirement of the drive motor. It should be noted, however, that the power input to an electric motor is greater than would be indicated by the horsepower because of the inefficiency of the motor. A 10 percent loss in efficiency at the motor becomes a significant factor in economic calculations.

It is important to study the pump curves, and specify a motor size that will be able to handle any possible power requirement, from shut-off to run-out. There are some pump curves which feature a power curve which is straight or drooping at both ends. In such case the selection of motor horsepower is simple, as the driver need only be able to handle the highest value on the curve. Overall efficiency of a pump is dependent on many of the pump and system parameters discussed in following pages. Efficiency curves are normally based on tests conducted by the manufacturer, and are useful for operating analysis. A pump should be selected for operating efficiency to be somewhat to the left of the best efficiency point. Since pumps rarely operate at design point, and departure of normal operation toward the right will result in better efficiency and minimal overload problems, although this is by no means a universal opinion!

An important factor in determining bhp is the head developed by the pump, called total head or total discharge head, TH or TDH. Referring to Fig.   , the TDH may be calculated as follows:

$$TDH = \frac{(P_2 - P_1)^{2.31}}{Sp.\ Gr.} + Z_2 + H_{fs} + H_{fd} \tag{7}$$

P_2 = Pressure on liquid surface in discharge tank in psia. (See P_1 above for more detail.)

P_1 = Pressure on liquid surface in suction tank in psia (See P_1 above.)

Z_2 = Height in feet of liquid surface in discharge tank, above surface in suction tank.

H_{fs} = Friction loss in feet in suction line, as above.

H_{fd} = Friction loss in feet in discharge line including exit loss from pipe into tank.

The fact that a large motor may be applied for nonoverload operation does not affect pump efficiency since pump shaft bhp usage will remain the same.

Care should be taken when applying pump curve hp for an efficiency approximation that the illustrated hp does not include motor service factor. Motor service factor is the motor manufacturer's safety factor and "down rates" the actual motor hp ability to provide against low-voltage operation, etc. On standard open-type a-c motor of 3 hp and over, a service factor of 1.15 is applied. The service factors increase with decreased hp as illustrated in the following table:

Horsepower	Service Factor
1/20	1.4
1/12	1.4
1/8	1.4
1/6	1.35
1/4	1.35
1/3	1.35
1/2	1.25
1	1.25
1½	1.20
2	1.20
3 & up	1.15

Standard Open A-C Motors Only (40°C Rise)

There are a few rules that generally apply to rotary pumps and perhaps should be mentioned at this point.

- Slip capacity is a function of viscosity and discharge pressure only and not affected by speed. Slip capacity is defined as the amount of fluid leaking from the pump discharge to pump suction through the internal pump clearances. Assuming no change in clearances, this leakage will only be affected, for all practical purposes, by the discharge pressure and viscosity of the fluid being pumped.

- Capacity is proportional to speed at 0 psig discharge pressure.

- Slip capacity is a logarithmic function of viscosity at constant discharge pressure.

- Friction horsepower is a logarithmic function of viscosity at constant speed.

A curve should be furnished indicating capacity, brake horse-power, and mechanical efficiency plotted against discharge pressure, for both maximum and minimum viscosities specified. Curves of capacity and brake horsepower to originate at the abscissa.

Index

411